Innopolis University - From Zero to Hero

Manuel Mazzara • Giancarlo Succi •
Alexander Tormasov

Innopolis University - From Zero to Hero

Ten Years of Challenges and Victories

 Springer

Manuel Mazzara
Innopolis University
Innopolis, Tatarstan Republic, Russia

Giancarlo Succi
Innopolis University
Innopolis, Tatarstan Republic, Russia

Alexander Tormasov
Innopolis University
Innopolis, Tatarstan Republic, Russia

ISBN 978-3-030-98598-1   ISBN 978-3-030-98599-8   (eBook)
https://doi.org/10.1007/978-3-030-98599-8

This Springer imprint is published by the registered company Springer Nature Switzerland AG
The registered company address is: Gewerbestrasse 11, 6330 Cham, Switzerland

"How far that little candle throws his beams!
So shines a good deed in a weary world."
William Shakespeare
The Merchant of Venice

To all our dear colleagues, now that we leave 10 years of collective growth behind us, while we keep smoothing the stone and improving ourselves, as individuals and as a community. No man is an island.

"Never let the future disturb you. You will meet it, if you have to, with the same weapons of reason which today arm you against the present."
  Marcus Aurelius
  Meditations

Preface

Now that Innopolis University is in its tenth year and reached a student population of about 1000, with 400 employees (of which about 25 are faculty members) and teaching and research staff up to 200 employees, it is time to look at the past and at the future. Plans for growth are there—planning to reach a student population of 2000 within a few years—and tradition is behind us and with us. We are an established organization reaching the phase of maturity. The aim for the year 2030 is to transform the university into a multidimensional university, still focused on IT, but not just related to pure computer science, including possible synergistic domains of knowledge. For us, excellence is the worldwide recognition of competence in the target research areas, the attractiveness of excellent students ready to invest their life and their wealth in the education programs, and the high demand of research and consulting support from the industry. This is what we are looking for; this is where we will concentrate our creative energies in the incoming year. We hope the readers will be involved in the narration of our beginning and our present. We hope they will be excited about our future. We hope they will become part of this wheel that is spinning faster and faster. Welcome to our ecosystem!

Innopolis, Russia   Manuel Mazzara
Innopolis, Russia   Giancarlo Succi
Innopolis, Russia   Alexander Tormasov
February 2022

Acknowledgments

Each of the authors has individual and collective acknowledgments, people without which the entire enterprise would have been absolutely impossible. We, all together, thank Innopolis University for generously funding and supporting this endeavor. In particular, all the top management in the person of our director Kirill Scmenikhin and our provosts Evgenii Bobrov and Iskander Bariev. We are extremely grateful to all the administrative staff. Unfortunately, a full list cannot be provided here. Just a few names: Timur Tsiunchuk and his team for the continuous support to our research and teaching work and for offering part of his vast photographic archive for this book; Ekaterina Protsko and all his band, for leading the effort of selecting students to admit to the studies, the leading force behind the success of the enterprise; Petr Zhdanov for the continuous support in the internalization process; and Dinara Valieva, who supported the work helping with numerous details of the process that are invisible to the final reader, but that were necessary to safely and successfully complete the book.

Manuel Mazzara personally thanks his wife Inna, his son Alexandr-Emanuele, and his mother, Maria Teresa, for contributing to the energetic environment: calm and motivation, practical support, all necessary to find the balance and the concentration, and the time to write.

Prelude

When I was asked in 2012 to participate in the process of creating the university as rector and founder, my ideas about what to do seemed to be quite different from my current understanding. This book, which is offered to you, attempts to describe the process and the components of the university that are now in place. Here, instead, I would like to point out some of the "hidden" parts of this process. **Why were some non-obvious decisions taken?**

Numerous decisions made during the early stages still have a big impact on the present and the future of the university. They constituted the basis for the development of the university. The fewer mistakes are made at the beginning, the easier life becomes and the higher the probability to fully achieve the initial goals. I have to admit that some of those decisions may appear obscure to the external observer. The decision of teaching all students in English is one of them.

While this is quite natural for most international universities, several observers believed that doing this in Russia would have been just a pointless attempt to stand out from the pack, neglecting all the tradition of higher education. As far as I know, this was the first such attempt in Russia, in particular concerning a bachelor degree. For master programs, selecting students among English-speaking applicants is an option. However, for bachelor degrees, and in particular for students in their first year who have just graduated from school, the challenge of selecting applicants with good language proficiency can raise some concerns. We remembered the experience of Soviet times, when the number of school graduates with a satisfactory level of proficiency was extremely low. Finally, at Innopolis University, we required an upper-intermediate level for the freshmen. This was necessary to allow students to interact with teachers and with each other, solve problems, and write reports.

While evaluating potential teachers for our programs, we found out quite interesting facts. For example, looking at the world's leading scientists in physics, we observed that about 20–30% of them were affiliated with Russia or with the former Soviet Union. In a situation like this, it is not particularly difficult to find high-level professors in the relevant field able to teach in Russian language. When we started to test the same hypothesis in the field of computer science, we realized that the scenario was very different: among the 400–600 most cited scientists in computer science, only 2.5 were in any way related to Russia or the former Soviet Union.

It was indeed very difficult to look for highly cited specialists and world-class teachers in Russia. Naturally in Russia, there are a fairly large number of specialists who do not get into this game for various reasons: different objectives and incentives, focus on Russian-language journals, or other difficulties associated with English-language publications. However, even these people were settled at some place, and persuading them to come to a completely new place in a clean field did not seem very likely. We had therefore to look at the international market and focus on the flow of teachers from abroad. Once this was realized, the decision of using English as a language of instruction came naturally. Such a decision is also relevant in terms of competition for world university rankings. Essential conditions to enter such rankings are indeed global recognition and publication activities. Achieving this would be actually impossible when teaching in Russian and publishing in Russian journals and conferences.

Once it was decided that the language of instruction should have been English, and that international teachers should have been hired, another turning point emerged: at what point should we switch students to English-taught courses? We studied the experience of other countries with similar problems, for example, the experience of Korea, where teaching begins in Korean, then from the second year students start to be introduced to separate courses in English, and the third and fourth years are already conducted entirely in English. From my point of view, it looked like chopping off the tail in small slices. That is a painful process: better to do everything at once. Therefore, we decided to start from the very first year of study.

At the time of writing, about one-third of our students are from foreign countries. We have created an entirely English-speaking environment in line with global trends and requirements related to the field of computer science. The language of instruction is just one example of the problems we had to solve. There were other issues that we had to deal with:

- What should an institution of higher education do when it is starting from scratch?
- How do you draw attention to the organization?
- How do you solve the chicken and egg problem when no one goes to a new place because no one is there? This applied to the city as well as to the university itself.
- How do you set the bar high enough?

The defined answers were non-trivial, especially when observed from the outside. This book tries to answer some of these questions.

I would also like to point out that we intended to allocate quite a large area for potential robotics labs since in 2012 the robotics field was quite promising. This required equipment and a place to keep it. The university building was then equipped with a suitable area. We had a hall with a 5-ton monorail crane, large ceilings, and space that was virtually untapped for the first 5 years. Later on, it turned out that all the unmanned systems of various sizes, and other systems like cable 3D printers, can fit and exploit this hall.

Creating a new university is a rather large and complex task. We have to understand how to get a model for its existence, including a financial model that will be self-sufficient and self-sustaining after an initial period of time. It is clearly necessary to ensure its survival in the long run. How to attract students? How to get teachers to match the level of students? This cannot be done by an immediate action. You cannot recruit 5000 students on the first course, launch their training straight away, and achieve good results. For these reasons, we also opted for a slow and gradual development.

One of the first decisions in this sense was licensing the Carnegie Mellon University master program in Software Engineering. Theoretically, we could have developed a similar program ourselves from scratch. However, we sensibly chose to transfer an existing program, despite the potential difficulties with adaptation and the relatively high price. Our decision was like the decision of any commercial company whether to develop some competence in-house or to buy it (license) from someone else: the second option gives a significant time saving—it is called a jump start. I think that if we had chosen the scheme of creating the program from zero on our own, we might not have been able to wait long enough for quality results (the expectations from the university were high; we would not have been allowed to wait too long for quality graduates). We started by sending students to actually study at Carnegie Mellon University and then went through a phase of gradually transferring their courses to us. Our next batch of students was twice as large: half went there, and the other half studied the same program with the same teachers at Innopolis. After some time, we fully transferred this program and found out that such complex programs cannot be simply adapted to Russian conditions, because they are oriented to the American market.

One of the specificities of the context at Carnegie Mellon University was that some applicants were participating in the program not with the self-sufficient goal of gaining new knowledge, but instead as a springboard for entering the US job market. Our specifics were different—we had to adapt to the state of the Russian economy and the recruitment of graduates into Russian companies. This allowed us to make such an updated version of the program over time. As will be discussed in the book, the program in software engineering was not the only one developed with external help. While with Carnegie Mellon University we licensed the entire program, with the University of Amsterdam we created programs together based on their approach and their technology, but adjusted to our needs.

Why have we also tried to shorten the educational programs? The specificity of Western fee-paying education is essentially that graduates receive quite high salaries after graduation. While they are studying, the money is deducted from their personal finances. Therefore, students have a natural desire to shorten the study period as much as possible since this is the time of money loss and without earnings. Accordingly, there is a tendency in the world to compress, even for bachelor programs, from 4 years to 3. As far as I know, in Great Britain several higher educational establishments have also switched to a 3-year bachelor program and a 1-year master program. We have also tried to reflect this situation, although for the majority of Russian universities, the necessity of a shorter regime did not arise. On the contrary, everybody complains about not having enough time and about the students being overloaded.

At Innopolis University, we accept the necessary balance between the complexity of the training and industry needs. Adapting to the market is also quite a complicated process to understand. It is easy to make some mistakes which would further complicate the integration with industry. There are very few universities in the world that would be as closely integrated into the industry as Innopolis University had to be. There are of course specialized universities attached to organizations, but they are rigidly tied to that particular company. Very often universities, especially in Russia, suffer from the fact that their graduate specialists are somewhere "in the air" and caring only about their grades and, at the same time, their destiny does not have much influence on the policy and life of the university. We, on the other hand, had to monitor actively whether our graduates interact with industry since a peculiarity of our financial development was that a significant share of our budget was provided by industry. Ça va sans dire that companies would have not supported the university if we had not met their needs.

Another important issue that was addressed was balancing fundamental education and practical skills necessary to operate inside companies. IT companies place more emphasis on practical communication skills. For example, if a company puts their HR director in contact with a university, they usually have a planning horizon limited to a few years, and their focus is on filling immediately positions with graduates having practical skills. These companies do not look more than a couple of years ahead because they know that in 3 years they will radically change their status: their top management will change, they will be either acquired or bankrupt, or they will grow so big that old ideas will be no longer relevant. This context imposes its own specifics to the interaction with industry.

The modern development of IT technologies, especially computer science and artificial intelligence, requires fundamental scientific knowledge also from an average engineer up to publications in leading scientific journals and participation in conferences. We have tried to take all this into account in the creation of our university. So far, we are only summing up the first results. We are approaching the first decade since the creation of our organization. We believe that we have much to be proud of: the individual achievements, the quality of our students, the position in the national rankings, and our presence at the international level. All of this makes us believe that our experience has resulted in a qualitative result, despite the mistakes, the problems, and the changes in the external environment, which sometimes dramatically changed the vector of development.

We could discuss all these questions further, but the reader will find the answer in the book. I would also like to express my personal gratitude to all the colleagues with whom I have worked both within the university and in Tatarstan, including the Government of Tatarstan, and of course in all of Russia and abroad. Their contribution cannot be underestimated. I will not try to list here all those who influenced and had a hand in the creation of the university, because it would take another whole book.

Innopolis, Russia   Alexander Tormasov
February 2022

Contents

Part I
The Beginning

The first part of the book focuses on the early days of the project: Innopolis University before the construction of Innopolis city. We present the concept, the motivation, and the objectives for the development of a new university. We share memories of the initial set-up of the university in the center of Kazan and the period of the construction of the city. We confronted the challenges of scaling from small to big, of adjusting to the new premises, and of setting up the research spaces. All this was done with incredible enthusiasm in a very short time, with sleepless nights, and, at times, with some improvisation. This was fun and came out just great.

Chapter 1
The Arcadia of Innopolis

One aspect that strikes foreign visitors when they first come to the campus is the impressive real estate. A common impression of new visitors, since the early days, is that it had to exist for at least 20 or 30 years, while indeed it is only a few years old. What happened before the construction of the campus? Innopolis already existed as a small office on two floors in a business center in Kazan downtown. This chapter presents some memories from this Arcadia where everything had a beginning.

1.1   The Early Days of Innopolis

Arcadia in literature is a physical or metaphorical place associated to harmony with nature. The term is derived from a Greek province and over centuries developed into a poetic idea of an idyllic vision of unspoiled wilderness. We like to adopt here this concept when referring to the early days of the vision of Innopolis University and the vast land where the city was later built, at least how it looks like in summertime. The reader will give us this concession at the opening of the curtain. Figure 1.1 shows one of the many pictorial representations of such an ideal idyllic place and time.

In practice, when one of the authors (M. Mazzara) visited Kazan for the first time, in December 2013, the site where the city of Innopolis has now settled was only a vast land covered by snow. Literally, a greenfield in the hills of Tatarstan, or better to say a *whitefield*.

Great plans and visions were on the paper, but it would have been hard to predict that such a vision could actually have been implemented to such an extent in this short time. Even for the most visionary and optimistic person. But indeed it has been. The city is there; you can see it already coming when you cross the Volga bridge on the M7 highway. The major Technopark is visible on the right, on top of the hill. This, together with the beautiful Volga landscape, creates a feeling of

© The Author(s) 2022
M. Mazzara et al., *Innopolis University - From Zero to Hero*,
https://doi.org/10.1007/978-3-030-98599-8_1

Fig. 1.1 Dream of Arcadia, Thomas Cole (1838). Public domain, via Wikimedia Commons

anticipation on the new visitors. You can share these sensations when you bring newcomers to the city by car, from the airport. The expectation is growing little by little all along the trip and the landscape transformation: from a modern hi-tech city like Kazan, to the idyllic countryside of the region, to then turn again into a modern scenario where you would not expect it.

In the early days (2013–2015), the first operational center of the university was based in Kazan, where a small group of about 30 people started to develop the concept and the early educational programs: the first idea of bachelor program and the inherited Software Engineering master program developed together with CMU university. The first bunch of 14 master students spent 1 year in Pittsburgh, USA (2013–2014), in order to graduate in the program and being trained for then delivering the program in Kazan, as local teaching assistant. Some professors were hired and sent for training to the USA too. This was a massive logistical and financial operation that allowed the university to start up in an effective manner, getting also the attention of the national media and worldwide.

In the small operational center of Kazan, the core of the first bachelor program was designed with long discussions in the only meeting room initially present there. At the time, one of the key partners of the university was ETH, the polytechnic school of Zurich, and one of the visiting professors was Bertrand Meyer,[1] presently an emeritus professor of ETH and still visiting professor of Innopolis. Professor Meyer was instrumental in the startup phase of the organization, as well as David

---

[1] https://en.wikipedia.org/wiki/Bertrand_Meyer.

Fig. 1.2 First international meeting of SE lab

Vernon[2] who was also visiting professor and the main consultant of the provost of the time, Tanya Stanko. These two prominent figures had a strong influence in the creation of the early BS and MS programs.

Given the presence of Meyer, and as a natural consequence, while the Master in Software Engineering was inspired by CMU (and later the Master in Network and Security was designed together with the University of Amsterdam), the bachelor program got many elements of inspiration from the Swiss Program.

Figure 1.2 shows one of the early meetings of the Software Engineering lab in November 2014. Figure 1.3 represents a genuine discussion between Bertrand Meyer and Salvatore Distefano, another one of the early visiting professors (now at the University of Messina), about the application of design by contract [11] to non-functional requirements.

During those months, many pedagogical discussions had to be put on the table, with frictions and discussion sometime, as it can happen in every human environment. We were trying to develop a program that would offer all the classic foundations of the program presented by top universities, but at the same time offering some element of uniqueness, some kind of cherry on top of the pie. The solution to this problem appeared relatively early studying the program of ETH. However, it was not exempt from troubles.

---

[2] https://en.wikipedia.org/wiki/David_Vernon_(professor).

Fig. 1.3 Dinner discussion

It is worth mentioning the decision about this point of uniqueness among all the other pedagogical issue. This choice costed an enormous amount of time of discussion and sometime the (we believe, unjustified) dissatisfaction of students. Eiffel was chosen as the first programming language to be thought to our students, the introductory language. This was the cherry on the pie we were looking for, this element of uniqueness. Among other supporting elements, this was exactly the same choice made by ETH (at least until 2016).

1.2 Eiffel as First Programming Language

The choice of Eiffel and design by contract as programming and methodological tools for first year bachelor has been long discussed and at times opposed inside the university. Ex-post, we can confidently affirm that it was the right choice. However, it is worth here spending some time to motivate the decision.

Which programming language is better to start studying for the beginners? There is no single correct answer to this question, no common agreement on this, and debate in some cases is vibrant. There is no global consent between programmers, teachers, academics, and business people. The reason is simple: the question is incorrectly formulated. For comparison, imagine that the question is not about programming languages but about human languages: what is the best foreign language to learn? Clearly, there is no single correct answer. The answer depends on such factors as the student's mother tongue, his cultural background, his objectives and motivation (professional or leisure?), and others.

The same goes for programming languages: no single answer exists. In general, it is easier to answer to the opposite question: "What programming language is better not to start with?". Experience has shown that teaching a specific language from scratch in order to satisfy a specific and urgent need rarely brings the individual to develop into a great professional. There are, for example, many cases of individuals who improvised themselves as Visual Basic programmers from scratch or moved from FORTRAN or COBOL to Java because of some local business need. Even if they managed to patch the immediate emergency, they rarely developed into great professionals, because they lacked the correct mindset and basic skills required by an experienced professional. Sometimes, such an emergency is inevitable, but developing a quality curriculum for a top-level university requires more care.

Worldwide, examples of good pedagogical approaches for programming are not missing. There are a few preliminary considerations. First, what programming paradigm we want to use? There is a general tendency to prefer the object-oriented programming (OOP) paradigm as starting point since it has proved to develop the right abstraction skills and methodological attitude. This approach, however, is not without its critics: some believe that object orientation may deal too much with design and interface aspects and not enough to algorithmic details and imperative flow structure. According to this view, procedural programming would be better to start, while object orientation should be introduced in advanced courses. Of course, this depends on how the course is taught, but the concern is serious. The school of thought privileging OOP usually concentrates on languages like Java or C# in order to take into account business demand. The school of procedural programming sometime concentrates on purely academic languages, like Pascal, with the benefit of simplicity or widespread languages, like C, offering broader flexibility (and related complexity). There are other paradigms too, for example, the Functional (Lisp, ML, Scala, etc.), which has attracted renewed interest in recent years, and Logic (Prolog).

The second general observation is that any serious computer scientist or software engineer will learn many languages over the course of his career. In particular, almost everyone will learn one or more of the dominant languages such as, today, Java, C#, C, C++, or Python. So the choice of the first programming language is not exclusive of others; rather, it is a preparation for others and should emphasize the development of the skills needed to learn programming. (In fact, an increasing number of students have done some Java or other programming before they even join the university program.)

A related question is: "how much emphasis we want to give on formal reasoning and correctness by construction?". University is the ideal time of life for learning new things and, at the same time, build the foundations of one's knowledge and mindset. Establishing a broad and deep basis is also the best way to make sure that students not only receive sufficient initial training to obtain a first job but acquire the extensive long-term intellectual skills to pursue a successive career over several decades: the technologies will change, particularly in such a quickly evolving field as Information Technology, but the principles will remain. As result, a broad school of thought supports the idea that the introductory programming course and the first programming language should emphasize computer science foundations and formal reasoning at the time of learning the first language, to strengthen a mindset leading to the development of better software. Our experience of teaching Eiffel in Introduction to Programming at IU implies the choice of OOP and an affirmative answer to the second question, and still Eiffel is only one of the possible choices to which these decisions lead. This path is not free of controversy. The experience inherited from ETH Zürich is positive, and the course was well taken by students. We aim at repeating the success in different contexts, though an adaptation phase is necessary and benefits of the approach may not appear as immediate.

1.3 Innopolis Today

Innopolis University is now approaching its tenth year and reached a student population of almost 1000 with 400 employees of which about 25 are faculty members and teaching and research staff up to 200 people. There are plans to reach 2000 students in the incoming years. The university is an integral part of the city, which has now two operational technoparks and two more under construction (at the moment of writing). The flow of students between the university and the companies of technoparks is regular and growing, fulfilling what was, since the beginning, one of the key measures of success: cooperation with industry and supply of highly qualified professionals.

The university has an extensively developed network of international institutions collaborating under different formats: student exchange, Erasmus +, visiting professors, joint PhD supervision, joint projects, and summer internships offline and online. One of the collaborative projects has seen as partners CERN and Newcastle University [2, 4], and collaborative PhD supervisions involve several universities including Toulouse, Nice, and Nantes in France, the University of Southern Denmark, the University of Messina in Italy, and the Brno University of Technology in the Czech Republic. All these activities dramatically supported the internationalization of the project. In turn, the growth and internationalization of the university also helped the development of the city itself bringing professionals, students, and talents from abroad, once the city attracted attention worldwide.

With the prerogative of a selective education free of charge and a constant attention to the internationalization of teaching and research, Innopolis is aiming at exploiting the benefit of the global trends without suffering the risks, such as excessive emphasis on the market aspects of education and the death run to global rankings which would put at risk the fundamental academic values [8].

During the pandemic, the model of Innopolis University did not suffer particularly as those of other countries. Not relying financially on the stream of foreign students, the struggle of the period was moderate and contained, thanks to the ability of management and employees, who all put extra effort to keep things going.

1.4 Innopolis Tomorrow?

While we are writing, Innopolis city is under continuous development and so is the university. Two of the four technoparks presented in the master plan are already operational, and two are under construction. Two new student dorms are also under construction, and they will double the capacity of the four existing buildings. New townhouses and apartments will also be soon available. The plan is, in the incoming years, to increase the capacity from about 1000 to about 2000 students. This significant growth requires an expansion in terms of staff and in particular faculty. The challenges for faculty growth are now different than in 2013, but not easier. The pandemic which started in 2020 changed the market and the perception of people and to some extent their ability or willingness to move. At the same time, many universities which relied mostly on foreign students for their income (e.g., the UK and Australia) are experiencing financial troubles. Innopolis University had a different financial model since the beginning that now appears to be successful and may have an advantage out of the global situation.

In the remaining chapters of this book, we will analyze some of these challenges, in terms of professors recruitment, faculty growth, student enrolment, and university development. We will look at the present and the future of the university and, as a consequence, of the city.

Chapter 2
Early Days and Further Development

We present here the key people supporting the university development, the legacy of Russian educational system combined with the novelty of the project. The past and the future live in Innopolis in a unique blend. As a private university, there are also challenges.

2.1 The Foundation

Innopolis University has been registered on December 10, 2012. At the early stage, it operated in Kazan downtown and moved to Innopolis city during the period April-September 2015, after the completion of the new campus.

The foundation of the university was announced in February 2012, when negotiations were held with Carnegie Mellon University on the creation of an IT personnel training center in Tatarstan. In July 2012, the president of Tatarstan Rustam Minnikhanov met with Gil Taran, head of the iCarnegie Global Learning, subsidiary of Carnegie Mellon University. The parties agreed to create a new IT university.

As administrative director of the university was appointed Dmitry Kondratyev, candidate of physical and mathematical sciences, entrepreneur, and founder of the network of educational centers for schoolchildren "Unium". As a rector was appointed one of the authors of this volume: Professor Alexander Tormasov, Doctor of Physical and Mathematical Sciences (Fig. 2.1).

On December 10, 2012, the initial legal registration as an autonomous non-profit organization of higher education was completed. On December 21, 2012, iCarnegie and "Innopolis University" signed a memorandum of understanding.

At the end of 2013, Russian Prime Minister Dmitry Medvedev approved the construction of the Innopolis University in the city of Innopolis.

© The Author(s) 2022
M. Mazzara et al., *Innopolis University - From Zero to Hero*,
https://doi.org/10.1007/978-3-030-98599-8_2

Fig. 2.1 The rector of the university

2.2   The Startup with CMU University

The specialists of the Carnegie Mellon University department studied the Russian IT industry and the quality of IT personnel training. The results of the research were later exploited to develop the educational concept. Specialists from the American university were involved in the development of the infrastructure of the university and the creation of educational programs.

In April 2013, Innopolis University announced a competitive recruitment. There were about 700 applications, and only 14 candidates were chosen to attend the Software Engineering program at Carnegie Mellon University for 1 year.

During the same year, 2013, the STEM (Science, Technology, Engineering, Mathematics) center of the university was opened in Kazan. Here, students in grades 6–11 were taught robotics, mathematics, physics, programming, and English.

2.3   The First Cohort of Bachelors

In February 2014, the university announced the recruitment of bachelor students. Candidates in years from third to fifth were considered from local and foreign universities.

The plan was to enroll about 40–50 students, with only 26 actually selected out of about 300 applications. They began their studies in the summer of 2014 at the Kazan site of the university. In August 2014, Innopolis University received a license for higher education and postgraduate programs. In May 2015, the university signed a cooperation agreement with CERN.

2.4 The New Campus

The city was inaugurated in June 2015. During 2015, according to the results of the selections, which took place from May to July, more than 350 students from 45 regions of Russia and 10 foreign countries were accepted. In August 2015, the training of students began in the buildings of the university in the city of Innopolis. Figure 2.2 shows a night picture of the university building, while Fig. 2.3 presents a moment of work for early movers into the university building (Spring 2015).

Figure 2.4 represents an introductory event held at the construction site in August 2014.

On October 1, 2015, Kirill Semenikhin, a member of the board of directors of Microsoft in Russia, replaced Dmitry Kondratyev as director of Innopolis University.

Fig. 2.2 The university building

Fig. 2.3 Work of early movers into the university building

Fig. 2.4 The construction site

2.5 The Recruitment of Students

In 2016, the university accepted 313 new students from 46 constituent entities of the Russian Federation and 10 foreign countries; in 2017, 255 people from 42 constituent entities of the Russian Federation and 29 foreign countries were enrolled; and in 2018, there were 254 people from 38 regions of the Russian Federation and 33 foreign countries. In 2020/2021, 823 students study at the university.

2.6   The Competence Center: Robotics and Mechatronics

On May 10, 2017, the university announced winning a grant competition for the creation and development of the NTI Competence Center in the direction of "Technologies of Robotics and Mechatronics Components". In June 2018, the NTI Competence Center was officially opened in the direction of "Technologies of Robotics and Mechatronics Components".

The consortium of the center includes 56 partners among leading universities and academic institutions of the country, large industrial enterprises, and foreign partners: Sberbank, Aeroflot, Russian Railways, Gazprom Neft, RUSAL, KAMAZ, ITMO, FEFU, VolgGTU, and IzhSTU. In May 2019, the center's specialists, together with members of the consortium, presented a roadmap for the development of robotics and sensorics in Russia until 2024.

2.7   The Chief Data Officer Training Program

At the end of 2019, Innopolis University, within the framework of the federal project "Human Resources for the Digital Economy", trained 2150 Russian civil servants under the "CDO (Chief Data Officer)-data-based management" program. The IT university conducted training for 41 CDO managers from the constituent entities of Russia with a trip to Singapore and Barcelona. Also, at the initiative of Tatarstan President Rustam Minnikhanov, 79 deputy ministers and municipal leaders of the republic passed the CDTO-Head of Digital Transformation program at the university.

2.8   Leading Research Center for the Digital Economy

In January 2020, Innopolis University received the status of a leading research center for the digital economy in the field of blockchain winning a competition from the Ministry of Digital Development, Communications and Mass Media of the Russian Federation. By 2021, the center's specialists plan to create a fully verified blockchain platform. The project partner is Aeroflot.

2.9   The Rankings

The university reached the 87th place in the ranking of Institutions Active in Technical Games Research becoming the only Russian university in this list. The ranking is compiled by Mark Nelson, professor of computer science at the American University in Washington.

In 2020, Innopolis University took the eighth place in the category "Joint international scientific publications" of the U-Multirank rating, founded with the financial assistance of the Erasmus + program of the European Commission. The Russian IT university became the first Russian university to enter the top 25 of this category. In total, the rating compilers evaluated 1700 universities from 92 countries in 10 categories.

2.10   2020 and Beyond

In December 2020, Innopolis University received a federal subsidy of 6.4 billion rubles. As a consequence of this, the Reference Educational Center and the Unified Methodological Center started their operations. Within this framework, the qualifications of teachers of higher and secondary vocational education will be upgraded, thanks to new programs for IT specialties and other subject areas. By 2024, the university should train 80 thousand Russian teachers, which is 30% of the total number of teaching staff in secondary and higher education in Russia.

In December 2020, Innopolis University created the first Institute of Artificial Intelligence in Russia, within the framework of which it combined existing laboratories and existing developments and research in the field of Artificial Intelligence.

Acknowledgement   The authors would like to thank Sergey Masyagin for helping with this chapter.

Part II
The People

The success of any enterprise is always mostly determined by a key factor: the people. In the case of Innopolis University, this means administration, faculty, and students. They all blend in the essential processes run by the university: teaching. The second part of the book describes the complexity behind hiring and developing the faculty, attracting students, designing teaching programs, and keeping a productive environment.

Chapter 3
Hiring and Developing an International Faculty

We discuss now the ideas behind the international faculty, the hiring process, the relocation of people, the training and coaching, and the growth in size of the faculty. We share how all the process bootstrapped and the challenges of starting from zero.

3.1 Booting the Recruitment Process

Professors select the universities to apply to and to join often based on the reputation of the organization, which may play a role higher than the salary, as the name of the institution on the business card of the professor constitutes a major plus when he/she performs private consulting or applies for research grants and funding for startups, which in turn are often a substantial part of the overall income. Moreover, the reputation of the current employer is a very important factor when a professor wants to change the job.

These factors themselves indeed explain the enormous difficulties that IU faced when it started the recruitment campaign for faculty members as early as 2013, just after its foundation, when still the city and the university were just plans on a paper.

The strategy was then to target mainly two classes of professors:

- *rising stars*, meaning young, freshly graduate professors, offering them a very attractive research pack, a personal lab with equipment and research assistants fully funded by the university for a substantial number of years,
- *old lions*, meaning well-established faculty members, also close to retirement, which could come either as permanent faculty members or visiting/adjunct and enjoy the possibility to have a new lab fully funded by the university where they could continue and extend their work, at the moment and also beyond retirement, plus the possibility to brand part of the university research and education with their personal research and educational choices.

© The Author(s) 2022
M. Mazzara et al., *Innopolis University - From Zero to Hero*,
https://doi.org/10.1007/978-3-030-98599-8_3

Moreover, the youngest of the *rising stars* were also offered the possibility to spend time in the existing lab of the *old lions*, in their original institutions, to promote cross-fertilization and speed up the amalgamation process.

Two additional key factors for both were the possibility to shape the curriculum anew, in a striking modern and innovative way, a situation unimaginable in established institutions and the availability of top-quality students. These students could have been educated based on the mentioned curriculum, and in a few years, they would have been ready to perform top-quality research, as then happened, as we can see from the results at the ACM student research competitions.

Initially, the contact with perspective professors was anyway difficult, mainly addressing potential faculty one by one by email after an extensive web search. This is how one of the authors of this book has been contacted in October 2013. The second round was done using LinkedIn, in particular the contacts of the people already hired.

Rising awareness about Innopolis in 2013 and 2014 was a difficult exercise. As soon as a core of specialist known in their areas was hired, the visibility of the university quickly boosted, and a regular hiring campaign via traditional channels (journals, conferences, mailing lists, etc.) become possible.

The establishment of a program of visiting professors from abroad also helped dramatically to this regard. At the moment, the faculty includes more than 20 members, of which a good share are foreigners.

Figure 3.1 shows the first moment in which the existing faculty visited the inside of the university building in February 2015.

Fig. 3.1 The first visit inside the university building, February 2015

3.2   Formalizing the Hiring Process

After the initial ad hoc, almost one-to-one phase, the faculty grew in size, and it became then a compelling issue to formalize the hiring process with clean and transparent rules, ready to handle also hundreds of applications. For this reason, an approach similar to the one of Google or Facebook was selected and includes the following steps.

The recruitment for permanent and adjunct faculty is conducted through the following phases:

1. First of all, as it is a common process, applicants are asked to send a cover letter explaining their motivations to join Innopolis University, a CV, a teaching statement, and a research statement; moreover, they are asked to have at least two research scholars to send a recommendation letter directly to the university.
2. A permanent hiring committee shortlist candidates on the basis of their application documents and the current information available online on scientific repositories, such as Scopus, dblp, Google Scholar, etc. This tasks involves also the deans and the institute heads of the perspective faculty and institute(s).
3. A person from HR with the support of a person with technical knowledge conducts a first online interview to test motivation and soft skills and some general background.
4. If the candidate is deemed suitable, then a second round of online interviews is conducted, in order by the deans and the institute heads of the perspective faculty and institute(s) and by two experts in the field identified by the permanent hiring committee; each of these interviews acts as a filter to the next.
5. On the basis of the results of the interview, the permanent hiring committee can take three possible decisions:

   a. to invite the candidate for an on-site interview;
   b. to defer the decision, in which case more interviews might be sought or simply step 5 can be iterated to gather better understanding of the candidate or of the need of the university;
   c. to reject the candidate with a rejection letter.

6. The on-site interview includes:

   a. a lecture (for students, faculty members, and research staff);
   b. a research seminar (for students and research staff);
   c. up to four individual interviews with faculty members;
   d. lunch and dinner, as/when appropriate, to verify further motivation and soft skills;
   e. one panel interview.

7. Feedback from all these parts of the on-site interview is collected, and, on such basis, the permanent hiring committee writes a synthetic recommendation for the appropriate hiring body.

We have found this approach very effective even if a bit long. Out of the people hired with this approach, practically none had any problem in getting adjusted to our working style and process. However, remember that not all candidates have to go through all the stages. If there is a doubt in any of the steps, the process is simply stopped. Moreover, we have empirically noted that we had a few candidates complaining for the length; however, they were candidates that eventually did not even reach the final stage of the process and were rejected in due course.

With the advent of the pandemics, the last stage of on-site lectures and talks has been moved online, indeed, missing the more convivial dining part.

3.3   Faculty Continuous Development

A number of activities have been established aimed at supporting new faculty during the initial period and all faculty along their careers at Innopolis University. These activities are integral part of the attractive package that is offered to faculty, and it is particularly precious for young hires.

3.3.1   Induction Workshop

During the first semester of working at Innopolis University, new faculty members attend a half-a-day workshop run by more senior colleagues aimed at introducing the working practice of the university, its operations, the vision, and the short-term and long-term goals, including the strategies to achieve them.

3.3.2   Teaching Improvement Programs

Innopolis University offers to its teaching staff the opportunity to attend on-site the Instructional Skills Workshop (ISW). The Instructional Skills Workshops (ISW) was founded in 1979 in British Colombia, Canada, to quickly give classroom instruction skills to engineers who had never had teachers' training. The process is an intensive introduction to teaching in a higher educational setting and focuses on practical in-class skillsets and course delivery.

This process was brought over to Russia via Innopolis University in 2017 and has since been the critical training unit for Innopolis Faculty at all ranks. Innopolis University's trainers are the regional "local representatives" of the program for Russia. Our first faculty received their trainer qualifications, from a founding member of the process, in 2018 meaning that we are entirely self-sufficient in the program.

Fig. 3.2 ISW graduation 2021

ISW is typically a 3-day event focusing on how to teach groups of students. It has been designed to enhance the teaching effectiveness of both new and experienced educators. A certificate of completion will be awarded to each participant. The participation is mandatory for junior teaching staff.

Figure 3.2 shows one of the most recent ISW graduation moments for some of our colleagues.

The ISW is a framework model of teaching practice. The format varies depending upon participant goals. The workshop begins with an overview of instruction. To be certified, an ISW must:

- Be a minimum of 24 h of in-classroom instruction (plus homework)
- Have a minimum of three peer-evaluated lessons
- Use an educational model of practice
- Engage with the process of the training

The participants of the ISW report that it has improved their relations with students and teaching qualities.

ISW at Innopolis

From the faculty side, Dr. Joseph Alexander Brown, to improve TA actions, started a series of workshop sessions with TAs. From the administration side, Oksana Zhirosh leads the creation of a needs assessment process. The findings of both processes showed the need for better faculty training in teaching, and it was to be a formalized process.

After this evaluation, many programs were examined for suitability. Finally, the ISW was chosen due to the following: (1) it is targeted to the goals of the participants, meaning that any faculty or staff member can join the session and gain from experience, (2) it is designed to focus on classroom skillsets and is application-based, (3) it is designed to promote a learner-centric model, and (4) the certification process is recognized about the globe and does not expire or require hefty fees to some third-party organization for "recertification" or "membership".[1]

David Tickner, one of the founders of the process, in 2017 ran the first ISW at Innopolis. This ISW was immediately followed by facilitator development workshops (FDW) and in 2018 with the trainer development workshop (TDW).

3.3.3   Individual Mentoring

Innopolis University offers a mentoring scheme for junior faculty or teaching assistants who wants to be assisted in their initial professional steps by a more senior colleague. The mentor is assigned by the Department of Education on request of the employee.

3.4   Faculty Evaluations

All instructors, professors, and professors of the practice at Innopolis University participate in a number of evaluations over the course of their careers pursuant to the following schedule:

1. All instructors, professors, and professors of the practice with teaching responsibilities have every course and laboratory they teach evaluated by students enrolled in the course or laboratory at the end of each semester;
2. All instructors, professors, and professors of the practice have their courses evaluated by faculty peers of the same or higher academic seniority;
3. Annually, all faculty members are evaluated by the dean of faculty, taking into account the evaluation of the director of the institute;

---

[1] https://innopolis.university/en/teachingexcellencecenter/trainings/.

4. In the last year of appointment, all faculty members are reviewed for reappointment;
5. When eligibility requirements are met, faculty members may apply for and be evaluated for promotion in academic rank.

As every process of this kind, especially in a new organization, everything is continuously in evolution, and the point described above may be changing in the future. Student and peer evaluation are described later in this chapter according to their current format. However, currently, there is an ongoing discussion on how peer reviews of classes should be conducted and whether it should be conducted at all.

3.4.1   Expected Conduct

All faculty members are expected to contribute to the missions of Teaching, Research, and Service as applicable to their individual appointment responsibilities and carry out these responsibilities in a professional manner. The exact ratio of effort and evaluation weight percentages in the areas of Teaching, Research, and Service is a function of the faculty member's appointment type, the skills of the faculty member, and the goals and long-term vision of the institute and university.

All members of the Innopolis University faculty are expected to follow a professional conduct and will be evaluated on the basis of performing assigned responsibilities with personal integrity; sympathy with and concern for colleagues, students, and others; as well as adherence to those Russian laws and university and institute policies, procedures, and regulations that have significance to the faculty member's professional performance and reputation.

3.4.2   Teaching

Faculty members with teaching responsibilities, regardless of academic rank, are responsible for teaching effectively by employing effective methods and approaches that facilitate student learning. Faculty members can demonstrate their teaching effectiveness through a number of different roles related to student learning. These include designing and delivering courses, directing undergraduate and graduate research, and mentoring students, whether formally or informally. Faculty can also demonstrate their commitment to student learning through involvement in curricular development, pedagogical innovation, and educational research.

Teaching faculty must demonstrate up-to-the-minute mastery of their subject matter, the ability to convey ideas in a clear and organized manner, and, when appropriate to the course subject, the ability to design engaging, hands-on active learning activities and assessments that improve the process of teaching and learning. They must also effectively use technology and other state-of-the-art teaching techniques in the classroom or laboratory, as well as show a willingness to learn and evolve as educators by participating regularly in faculty development opportunities and informing themselves about new teaching techniques, emerging educational technologies, etc.

Evaluation of teaching is based on a combination of assessments, including student course evaluations, syllabi review, grade distribution review, student letters, peer and director of institute classroom/lab observations, evaluation of teaching materials, teaching awards, and other evidence of contribution in this area.

3.4.3   Research Activity

Faculty-assigned research responsibilities are expected to engage in meaningful disciplinary research relevant to the research goals established by the university and the faculty member's institute, contribute to important discourse and discoveries in their field(s) of study, present at major disciplinary conferences, and publish or present in well-respected peer-reviewed and referred journals and publications. Research productivity will be measured foremost in terms of high-quality research.

Specific research performance measures include the following:

- Peer-Reviewed Publications and Presentations—It is expected that faculty members will present research findings at major disciplinary conferences and publish research results in well-respected peer-reviewed and referred journals, publications, or conferences. Expected performance as measured by the number and quality of publications and/or presentations varies by academic rank, appointment type, and discipline.
- Patents, Inventions, entrepreneurial and Other Discipline Advancement Activities—Research activities that contribute to the advancement of the discipline, including the development of any licensed software, patents, products, services, startup companies, and inventions; innovative and entrepreneurial proprietary, commercial, and professional service activities (e.g., transfer of technology to industry or other research organizations, etc.) related to the discipline; facilitation and management of large research programs; and research infrastructure development such as building new facilities and laboratories.
- Funding—Research applications, awards, and/or expenditures are expected to be at a level that is sufficient to support the activities of a research group comprised of a minimum of two MS/PhD/post-doctorate students, including its infrastructure in terms of equipment, supplies, conference travel support, etc. Faculty who participate in research center activities or government- or private-funded grants or contracts are expected to provide significant contributions to the activities, in terms of both collaboration on research and generation of funding, which will be given appropriate attribution when assessing research expenditures and output. Similarly, when specifically agreed to in advance by the university's academic administration, research funding from internal Innopolis University

sources will be given appropriate attribution when evaluating a faculty member's research funding, expenditures, and output.

- Students—It is expected that a successful research program will include participation of undergraduate students as well as graduate students and, in some cases, professional research assistants or research associates.

3.4.4   Service

Faculty members are expected to contribute to their departments and to the larger university community through participation in committees and other activities. Faculty should be involved in various institutes and/or university committees or other service activities such as performance of administrative functions, involvement with student groups, and activities. Chairing a major committee, directing a center, and the extent and effectiveness of participation are given additional consideration. Academic advising and outreach/recruiting activities are also elements of university service.

In addition, faculty members are expected to provide service to their professional discipline. External service brings recognition to the faculty member, his or her institute, the university, and the campus. The extent of service on professional committees, panels, or boards, as a journal editor and as a reviewer, is considered. Other examples of service to the professional discipline include but are not limited to serving as an appointed or elected officer of an academic or professional association; serving as an organizer or leader of local, national, or international conferences, workshops, panels, meetings, or summer schools in areas of professional competence; contributing time and expertise to further the work of a professional society or organization; promoting the image, prestige, and perceived value of a discipline or profession; participating in accreditation activities; refereeing manuscripts or grant proposals submitted to journals, professional meeting program committees, and funding organizations; and establishing professional or academic standards.

Consulting activities within the faculty member's discipline are also regarded as service to the profession as the university recognizes such activities to be a measure of the faculty member's value to society in the application of his or her academic knowledge and/or research findings. Volunteer activities in the public sector are good citizenship and are similarly recognized in the evaluation of service.

3.4.5   Student Evaluations

Student course evaluations are performed for all faculty with assigned teaching responsibilities, in all courses and laboratories, every semester. Student course evaluation forms are distributed near the end of the academic term to each student enrolled in the course/lab. The course/lab instructor is required to notify students in writing via the course syllabi and orally at the beginning and end of the academic term that completion of the student course evaluation form is mandatory and that student anonymity is maintained throughout the entirety of the course evaluation process.

3.4.6   Peer Evaluations

A faculty peer observation system is in place at Innopolis University for some faculty members with assigned teaching responsibilities. Initially, the peer evaluation was performed for every taught course and currently is performed only for new lecturers or new courses.

An experienced faculty member appointed by the dean's office performs peer evaluations annually during the first semester of the academic year. The role of senior faculty peer evaluator is to:

- Visit the class, laboratory, or other instructional setting in person or via a digital recording, live broadcast, or other electronic modality to evaluate the faculty member's teaching effectiveness if the faculty member has assigned teaching responsibilities; and
- Complete the classroom evaluation form and return it to the reviewee and dean of faculty within five business days of the evaluation.

Acknowledgement The authors would like to thank Sergey Masyagin and Joseph Brown for helping with this chapter.

Chapter 4
Curricula and Language of Instruction

In this chapter, we will cover the development, from scratch, of educational programs, the rationale behind the curricula, and the objectives and the bootstrapping of teaching, including the experience with CMU and the University of Amsterdam for the initial MS programs. We also discuss the manifesto of our intention for what concerns the language of instruction.

4.1 Bootstrapping

As mentioned in previous chapters, the first program launched by Innopolis University was a Master in Software Engineering developed in cooperation with Carnegie Mellon University (MSIT-SE). The first group of students were sent to the USS in fall 2013 to study for 1 year. Once graduated, this group served at the university as instructors. As the first director of the Software Engineering program was appointed Ales Zivkovic, in Fig. 4.1.

In April 2014, a competitive procedure was run to select 26 students to attend the first bachelor program of Innopolis University. Only students already attending the third year of other universities were admitted to the selection. The idea was to establish a pilot project able to graduate a group of students already after 1 year. It was a general pilot project.

The first bachelor program was launched then in August 2014. The starting date of the classes is unusual for Russia, and it is still unchanged at the moment of writing. It will be aligned with the national starting day (September 1) in the incoming year.

The unusual date was partly inspired by the academic schedule of Danish universities and partly organized to accommodate a winter break able to cover both Catholic and Orthodox Christmases, showing the international propulsion of the organization since the early days.

© The Author(s) 2022   29
M. Mazzara et al., *Innopolis University - From Zero to Hero*,
https://doi.org/10.1007/978-3-030-98599-8_4

Fig. 4.1 Director of MSIT-SE in a discussion of future trends in business education

4.2   The First Curriculum Without Tracks

In February-March 2015, it become clear the format of admission for the academic year 2015–2016. The new building was almost ready, and moving there would have required a fast scale-up in the number of students. Every weekend from March to July was dedicated to students' interviews. A massive work that allowed the university to enroll about 300 students.

All the faculty members in force in spring 2015 reached the building of the Innopolis University, still under construction, with a dedicated shuttle bus leaving Kazan on Sunday morning and coming back only in the evening. This period was particularly exhausting for the staff, but it led to a spectacular beginning of the academic year 2015–2016. Certain spaces of the new building were ready to use, with desks placed in a temporarily but functioning manner. Committee of three faculty members and assistants met every single students. Since then, the enrollment process was perfected, but it was notable since the beginning the high motivation and talent of the applicants and their adventurous attitude. The pioneering attitude started fading over the years since the new cohort of students progressively found better and better conditions and a stable process. However, the talent of our candidates makes of the enrollment experience always a nice activity to conduct. The advent of the pandemic in 2020 moved most of these interviews to online format.

In the cohort selected in spring 2015, there were first and third year bachelor and master students. Attracting them from other universities and enrolling to the third year was hard, but we could graduate a large cohort just after 2 years. The peculiarity and complexity of this year was in the fact that the three groups were

Fig. 4.2 Speech of Manuel
Mazzara at the graduation
ceremony of the first cohort
of graduates

taught the same courses. The teaching staff was tight, and the students numerous, and we had to have all the 300 students in the same room for lectures and separated only for labs. This optimization made the delivery difficult, but there was no other way to implement things in those pioneering times.

In 2017, we graduated the first group of bachelor students who lived in the city campus in the context of a majestic and very emotional ceremony. In Fig. 4.2, the speech of one of the authors (Mazzara) during the event that also signed the change of provost for education from Tanya Stanko to Sergey Masyagin. Current provost is Evgenii Bobrov, in charge since 2021.

4.3 The Organization in Tracks

The bachelor program was since 2017 organized in three tracks to be chosen by student during the third year of study: Software Engineering, Data Science, and Robotics. Only later, a fourth track in Computer Security was added.

In year 2019, also the first year of the bachelor program was organized in two tracks: Computer Science and Computer Engineering.

4.4 The Master Programs

Apart from the Software Engineering program, already mentioned, a master program in Robotics and one in Data Science were launched. During the period 2015–2016, some teaching staff spent the whole academic year at the University of Amsterdam to then launch in 2016 the master program in Security and Network

Engineering (SNE) in collaboration with the parental SNE program. This MS program is a 1-year-long intensive education and research together with industry, followed by 1 year of hands-on industrial experience. SNE selects the best candidates and covers competencies such as security engineering, security analysis, secure programming, penetration testing, and offensive security. So far, five batches of SNE have already been graduated and have been placed in the local industry. The value generated by the program is evident from the high-quality graduates and the fact that well-known companies are hiring these graduates. Since the launch of SNE program, the university had four master programs, and the organizational structure follows the same pattern. The structure of labs and institute will be detailed in the specific chapter.

The Master of Science degrees offered by the Faculty of Computer Science and Engineering at Innopolis University aims at providing its students a quality graduate education in both the theoretical and applied foundations of computer science. The goal is to train students through comprehensive educational programs, and research in collaboration with industry and government, to effectively apply this education to solve real-world problems and enhance the graduates' potential for high-quality, lifelong careers. The curricula are organized with courses distributed over 1 or 2 years. Such courses comprise a total of 120 ECTS (Europe Credit Transfer and Accumulation System), and they are performed mostly in English.

4.5 The Move to 3+1 Curriculum

In year 2021, a decision was made to move from a 4-year curriculum to a 3+1. Following the pattern already adopted for some MS program (e.g., the one in SE), it was decided that the fourth year of study would be implemented as 1-year-long internship in resident company, i.e., a full year of practice.

As a consequence of this decision, the original program originally unfolded over 4 years would be then developed over 3 years. This has been achieved with two key moves, plus work on the details: first, the addition of a summer semester where elective courses have been moved and delivered mostly online and second with the elimination of the pre-existing summer internships and having all the internship effort moved to the last year. A lot of work on the details was necessary, but the transition was relatively smooth. Starting from fall 2021, it has been implemented for all the first year BS students, and it will be extended to newcomers in the future. Those students enrolled before 2021 will complete instead the study according to the previous format.

4.6 Medium of Instruction

Innopolis University is committed to the highest international academic standards and offers the opportunity to study Computer Science and Engineering in English developing fluency and command of technical and non-technical vocabulary. This is made possible by the university's international community of staff and students, speaking several foreign languages, and in particular English, on a daily basis: students, professors, researchers, and administrative staff.

This commitment is reflected on using English as the language of instruction of Innopolis University. All classes, teaching, labs, assignments, exercises, and practical material are provided in English. English is expected to be used in the following situations:

- Frontal classes, including questions
- Frontal labs, including questions
- Class and labs material
- Technical and non-technical assignments
- Theses and projects
- Written and oral exams
- Group office hours
- Public chats related to university life
- Emails to students or colleagues
- Student's groups
- Any other document related to education

Having English as the official medium of instruction of Innopolis University does not prevent the use of other languages, and, in particular, Russian, in other situations where two or more people share a common language; thus, English is not necessarily used in situations like informal conversation (questions after classes, etc.), individual office hours, private emails, private chats, etc.

4.6.1 Rationale

Fluency in multiple languages reflects on the one hand a versatile mind and on the other hand the significant effort and dedication necessary to acquire such ability. Being multilingual is a notable skill, no matter what set of languages is in a student's portfolio; however, English is an important addition for every graduate. The importance of English language can be summarized in three major points:

- English as a communication means
- English as language of employment
- English as language of science

4.6.2   English as a Communication Means

English is the language of international communication; speaking English increases the chances to communicate virtually anywhere in the world in every field of knowledge. Fluency in this language allows graduate to travel, to encounter new cultures and ideas, and to grow as individual and professionals. It is the medium for exchanging technical knowledge as well as feelings and emotions.

4.6.3   English as Language of Employment

The opportunities available on the job market today requires graduates to manage a complex mix of soft and hard skills. Among these, language skills have a primarily importance, not only for international job market but also for seeking high-profile employment in the Russian Federation. All the major companies keep indeed continuous relations with abroad institution, and engineers, manager, salesmen, and every other highly qualified specialist need the command of at least one foreign language, typically English. English skills are just as desirable to employers in any country as they are to employers in English-speaking countries.

Fluency in English allows employees to attend and organize international business meetings. Where several languages are represented, the chances are that the meeting will be conducted in English. Employees with language skills will find themselves in a position of advantage and will be able attending relevant gathering, therefore having the opportunity to advance faster in the career ladder.

4.6.4   English as Language of Science

English is also the language of science, and being fluent allows graduates to continuously educate themselves accessing scientific literature and attending international conferences. English is essential to pursue a scientific career and to develop the necessary network able to develop your scientific research.

4.6.5   Commitment to Improve Language Skills

In order to gain admission to Innopolis University, it is necessary to provide evidence of specific language levels based on the specific program of enrollment. Over the study period, the English level is expected to grow. The university implements several ways to support students in their path, some of them being:

- English courses
- Communication courses
- Technical essays in English
- Lab internships
- International projects

4.6.6 Exceptions

Specific exceptions can be made by the Department of Education for courses that need to comply with laws and regulations of the Russian Federation (e.g., "safety"). The university commits to provide the opportunity to learn the material of these courses to all the international students.

In some cases, instructors who are fluent in Russian and possess the appropriate established terminology used in Russian industry and research can familiarize students with bilingual terminology that could give students a competitive advantage in the knowledge of modern terms used at the national and international levels.

Acknowledgement The authors would like to thank Adil Khan, Nursultan Askarbekuly, Rasheed Hussain, and Sergey Masyagin for helping with this chapter.

Chapter 5
Attracting the Best Students

Innopolis is now featuring the third place in Russia for quality of incoming students, according to the final grades of school state examination. The early days were different; how was this progress achieved? What has been the process?

5.1 Educational Model of Innopolis University

Innopolis University follows a model according to which all students are allowed to free education as long as they pass the selection process. At the time of writing, IU has a bachelor program in Computer Science and Engineering and four master programs:

- Software Engineering
- System and Network Engineering
- Data Science
- Robotics

From the academic year 2022–2023, the structure of the programs will be extended, and a growing number of admitted students are expected.

The first two of these programs were transferred, respectively, from Carnegie Mellon University (CMU) and the University of Amsterdam (UvA).

Both of them are quite unusual for typical Russian master's programs and designed for students with initial industrial experience. One of the main peculiarities is the so-called industrial project integrated in study plans. In the case of the master's program in Software Engineering (SE), students work for two semesters on a team project supervised by mentors from IU and industry. This practice was borrowed from Carnegie Mellon University, and the project presents real-industry tasks. In the project, students have the possibility to apply the knowledge obtained during the regular courses.

© The Author(s) 2022
M. Mazzara et al., *Innopolis University - From Zero to Hero*,
https://doi.org/10.1007/978-3-030-98599-8_5

The master program in System and Network Engineering has a similar approach transferred from the University of Amsterdam. In the frame of these programs, students have to work also on industrial projects, but different from the SE program, the length of the projects is just 2 months.

5.2 The Recruitment Strategy

IU had to perform its recruitment of national and international students always with a limited budget. Despite this, students showed an increasing interest for the university, both nationally and internationally.

The key elements that the university has adopted in terms of recruiting strategies could be summarized as follows:

- Use of major and capillary Internet channels;
- Word by mouth of satisfied current students;
- Incentives to current students to invite their talented friends;
- Recruiting campaigns in foreign countries (e.g., Italy in 2017);
- Program of visiting professors from abroad;
- Program of international internships;
- Student exchange (inbound and outbound);
- Erasmus+ with countries such Italy, France, the UK, Ireland, Turkey, Luxembourg, Denmark, etc.;
- Competitions and olympiads for students and schoolchildren.

Typically, the largest amount of students is coming from Asia and the Middle East. Only a small part is actually coming from America and Europe. This is due to the lack of awareness about IU in these continents. Dissemination activities from PR offices have been indeed still limited in these geographical areas.

The countries offering a large amount of students to the university are typically CIS countries, Egypt, and Pakistan. However, recent years have seen students enrolled also from Italy, Spain, and Latin America.

Figure 5.1 shows a moment of the selection of students.

5.3 Student Exchange

The exchange students and the international internships played an important role in placing Innopolis in the world academic map. In the framework of international programs, students of Innopolis University have a chance to visit world's leading university in the Computer Science sphere: National University of Singapore, KAIST, Seoul National University, Hong Kong University of Science and Technology, Polytechnic University of Milan, University La Sapienza in Rome,

Fig. 5.1 Student selection

University of Luxembourg, etc. Student exchange brings to the University, and to the participating individuals, several benefits, such as:

- An opportunity for students to extend their academic curriculum attending courses given in partner universities;
- International experience is a significant advantage from the employment perspectives;
- Experience of working in an international environment and new projects;
- Significantly increasing language skills;
- Broadening horizons and improving inter-cultural understanding.

On the way of setting up a completely new exchange network, IU faced several obstacles both on the inbound and outbound fronts. For what concerns outbound students, the difference in study plans and the technical recognition of received courses is a problem. To avoid this problem, since the beginning, IU looked for exchange cooperation only with universities that have a strong IT background and relevant degree programs focusing on IT, Computer Science, Data Science, Robotics, AI, Software Engineering, and Cyber Security. Additionally, IU helped to build up an appropriate individual study plan for each exchange student in order to make recognition as smooth as possible. Finally, all exchange partners provide English-taught courses at least at the MS level, so that students do not need to learn a new language to participate in the program.

On the inbound front, the major problem was related the young age of the university, which led to a limited international recognition by foreign students. That is why the university implemented several moves to change the situation:

- Each participant in an exchange was then used as an ambassador of IU to maximize recognition of the university in partner organizations;
- IU organized webinars for potential exchange students to learn more about the institution;
- Exchange coordinators disseminated various materials to partners;
- Starting from 2018, IU defined a scholarship for incoming students that covers transportation and accommodation.

In terms of financial support of mobility, IU was also involved in Erasmus+ scheme with partners from Turkey, the UK, Ireland, France, Luxembourg, Denmark, and Italy. This scheme provides funding for supporting not just student exchange but also staff mobility. Outgoing students have not faced serious financial problems, since they receive a monthly scholarship from IU on top of the Erasmus funding.

The fact that IU is an English-speaking university eases things in developing exchange programs: incoming students do not need to know Russian language and outgoing students already have a solid knowledge of English. Students at work are presented in Fig. 5.2.

Fig. 5.2 Students at work

5.4 Olympiads and Competitions

Olympiads and competitions have an important role in IU internationalization and student recruitment. For instance, in 2014, the university became a national partner of the World Robot Olympiad, which allowed to host the Russian stage of the event. This still brings hundreds of schoolchildren every year and provided a strong support for the training of the Russian national team.

By training the national team, IU was determinant in the achievement of top results. In 2018, the national team was first in the World Robot Olympiad and won 1/3 of all medals: 5 gold, 1 silver, and 2 bronze medals. Such results have positive influence on potential students, especially in the field of Robotics, and affect the choice of their future place of study.

5.5 Support for Students

Recruiting excellent students, placing them in competitions, and nurturing them are essential for the success of Innopolis. In order to continuously support the intellectual and emotional development of our most precious resources, the Students Affair Office is regularly organizing gathering, some sport-related, some entertainment-related, and others more academic-related, without any particular preference and unbalance in the assortment. An event organized to inform and support those considering a career in science is depicted in Fig. 5.3.

Fig. 5.3 An event to guide students to a career in science

Acknowledgement The authors would like to thank Ekaterina Protsko and Evgenii Serochudinov for helping with this chapter.

Chapter 6
The Research Environment and Our Values

After setting up the teaching process, research is fundamental to reach international recognition. Setting up from zero, an effective research process takes time. What assessment system should be created to motivate and incentivize researchers to perform their work at best? How to motivate professionals to join, stay, and work effectively? In this chapter, we analyze these questions and provide some answers based on our experience.

6.1 The Context

Innopolis University is a research active institution aiming at reaching visibility at the world level within the scientific community and placing high on the international rankings. In order to reach these objectives and increase the reputation, faculty engage in individual and community efforts to publish articles in high-impact venues following rigorous internationally recognized ethical standards. The building up of the environment is essential, but a good incentive system is a mechanism that cannot be missed in human environments.

6.2 The Values of the Environment

Hiring people from different cultures and academic systems, aggregating them, building a unified culture, and making them productive is not an easy task. Creating such a unified and motivated environment at the faculty level is then of fundamental importance to transmit the same values to young researchers, to PhD students, and, ultimately, to every new person who will be hired.

© The Author(s) 2022
M. Mazzara et al., *Innopolis University - From Zero to Hero*,
https://doi.org/10.1007/978-3-030-98599-8_6

To succeed in this enterprise, the first moves are critical, i.e., how the environment is built since the beginning and what practices are agreed and shared between colleagues. Respect and sharing are they key values here; however, it is also important to build processes that can keep the values even in the presence of turnover and when people, for professional or personal reasons, have to leave the university. The system should allow those who are leaving to leave a legacy that can be taken by the newcomers.

The construction of such a "human system" requires very long time, and its development is still ongoing. We cannot say that its stabilization has been fully realized, but we can see excellent signs that we are on a good track. The most visible evidence is the friendly cooperation between the faculty members, between the professors and the students, and between the students themselves (for not to forget the pleasant interactions with the administration).

The group is compact, and the processes are working smoothly, apart from some frictions that happen on a daily basis and are part of every human environment. However, every issue is settled before the sunset, and it does not drag along creating resentment. This is what we can personally experience and that shows a positive trend. We hope this can continue in the future, and we are working hard on this, with attention on the details.

6.2.1   Founding Principles

Faculty, student administration, and any other member of the Innopolis community should share some founding principles determining a code of conduct based on mutual respect as a pillar and a few others:

1. Respectful treatment of others
2. Individual freedom
3. Academic freedom
4. Intellectual integrity

Respectful Treatment of Others

Individuals' expectation of fair and respectful treatment by faculty and students applies not only to interactions with one another but also to administrators, staff, and others with whom they interact in their role as members of the academic community.

Individual Freedom

Individual freedom means that each individual has virtues and importance. All individuals have equal rights.

Academic Freedom

The notion of academic freedom lies at the very heart of the academic enterprise. In the "1940 Statement of Principles on Academic Freedom and Tenure", the American Association of University Professors (AAUP) states, "Academic freedom [...] applies to both teaching and research. Freedom in research is fundamental to the advancement of truth. Academic freedom in its teaching aspect is fundamental for the protection of the rights of the teacher in teaching and of the student to freedom in learning".[1]

Intellectual Integrity

Intellectual integrity involves using sound and ethical methods in the pursuit of knowledge, as well as embracing honesty in the dissemination of knowledge.

6.2.2   Code of Conduct

Here we cover in more details the main aspects of conduct that researchers at Innopolis University, and worldwide, may see challenged during their work. In any case, they should follow the founding principles presented above.

Plagiarism

Representing the ideas, words, or data of another person or persons as one's own constitutes plagiarism. Thus, a person's words, ideas, or data, whether published or unpublished, must be acknowledged as such.

Acknowledgment of Contributions

Acknowledgment of the contributions of others means appropriately recognizing and crediting those who have contributed to a scholarly work whether the work is a manuscript, exhibit, or performance. Depending on their contributions, such others, including students, may be deserving of credit ranging from acknowledgment in a footnote to co-authorship.

---

[1] https://www.aaup.org/report/1940-statement-principles-academic-freedom-and-tenure.

Data

Researchers must acknowledge the source(s) of their data and accurately describe the method by which their data was gathered. Moreover, the fabrication or falsification of data or results constitutes a violation of professional norms.

Conflict of Interest

Research funded by corporate sponsors potentially leads to a situation in which a conflict of interest may arise. Scholars must not let the source of their funding nor the sponsors' goals cloud their own professional and scientific judgments regarding their research.

Confidentiality

Researchers should protect the confidentiality of any professional or personal information about persons involved in research and scholarly activities.

6.3   The Incentive System

A research incentive system, in a context of experienced and highly qualified specialists, can work only if its nature is mostly endogenous, i.e., employees are self-motivated. To achieve that, the hiring process is fundamental. It is unthinkable to put in place a system of carrot and stick and expect that it can work in the long run. This is not how research works.

In the attempt to guide researchers, especially the juniors, the faculty council keeps up-to-date a list of publications ranked in two tiers (A and B). Venues indexed by the largest commercial databases such as Scopus of WoS are, in general, suitable targets too. The list maintained by the faculty has an advisory role and cannot limit the choice of the professors who ultimately are experts in a specific field and know their community. However, for career development, the faculty members are advised to concentrate the effort on journals and conferences with high impact and strictly avoid predatory publishers. The issue of avoiding predatory publishers may appear obvious to the experienced academics; however, we have seen that, without creating a suitable and informed environment, the problem may appear. In the end, without potential preys, such publishers would not exist. We understood that preys at risk are typically young researchers of organizations which failed to establish a proper culture and lacking role models.

Typically, a good selection of self-motivated individuals, an endogenous incentive system, and soft guidelines can lead the university to gain visibility, locally and worldwide, without the necessity of a costly, intense, and unpleasant monitoring system.

6.4   The Social Environment

A productive environment is based on trust and friendship. In order to build such environment, it is not only necessary to work together but to play together too. Innopolis University since the foundation has been always a catalyzer of moments of aggregation: from faculty international trips (e.g., in Italy in 2016, Fig. 6.1, and 2017 or in the neighboring regions of Russia) to dinners at the dean's place or in local restaurants and to playful situations organized to create and develop relationships in an informal context. Figure 6.2 shows one of these moments (May 2021).

Fig. 6.1 University trip to Italy in 2016

Fig. 6.2   Manuel Mazzara, Mirko Farina, and Giancarlo Succi: playing together

Part III
The Activities

Numerous activities determine the success of the university, and for each of them, complex processes had to be put in place. In this part, we will analyze in detail the creation of the organizational structure and the cooperation with industry. We will explain the rationale, the goals, the problems, and the solutions adopted.

Chapter 7
Faculty, Institutes, and Labs

Institute and labs are not only about spaces. It is necessary to create a structure where people can work with motivation and efficiency. We present here the structure and how it works.

7.1 The Importance of the Structure

An educational organization requires an effective structure in order to reach the planned objectives. Ultimately, the university structure is a management tool in the hand of the stakeholders that can simplify, or make more difficult, the daily operational activities. However, the university structure is also a way for the people themselves to reflect on their goals, and it is therefore an effective tool to build a system and orient the goals. With all this in mind, the overall structure of Innopolis University was built.

For what concerns the success of a university, people are certainly the most important part of the equation—in this case students, professors, and administrators—but excellent individuals poorly organized, structured, and managed will not be able to deliver optimal results being at the same time cost-effective. This is clear to every university management, and particularly important is to understand how to bootstrap the organizational structure since the early days.

The initial phase is important because the choices made during the startup will become legacy and will affect further moves. Once a specific structure, management style, and corporate culture have taken place, a revolution becomes difficult. It will be then necessary to modify, adjust, and correct in evolutionary terms, unless the stakeholders are ready for a major reset that involves costs, risks, and certainly losses. In general, new employees will adapt to the existing structure, culture, and mindset, with attached pros and cons.

© The Author(s) 2022
M. Mazzara et al., *Innopolis University - From Zero to Hero*,
https://doi.org/10.1007/978-3-030-98599-8_7

7.2   The Initial Structure

In terms of research divisions, the initial building in the center of Kazan included only two of them: the Software Engineering lab and the Robotics lab. Institutes were not present at the time; the small number of employees did not justify this level of the hierarchy.

Only when the staff moved to the new campus, summer 2015, a structure with three institutes was formed: Software Engineering, Information Systems, and Robotics. Each institute included a few labs; only a handful of faculty members were present at that time. At the first Faculty Council held in Kazan, a handful of professors participated, not because of absences, but due to the tiny size of the faculty reflecting the small number of students.

7.3   Faculty of Computer Science and Engineering

The Faculty of Computer Science and Engineering is the first one created inside the university from the legacy of the work done in the initial office in Kazan. As a novel institution, there has been a large turnover of faculty members and researchers over the years. It would be out of the scope of this chapter to discuss this evolution or to take a snapshot of every single moment. We present in Fig. 7.1 a photo of the faculty workshop at the end of year 2017.

7.3.1   Steps to Build a Faculty

As we mentioned, an organization is built around the people. A faculty is not an exception. If we observe the evolution, from the first group of people in Kazan to now, we can see that often the organizational structure was built around people. It was not people adapting to the organization, but the organization to evolve according to the need of people and their key values. Operationally, this process was performed bottom-up via series of meetings, workshops, and shared documents edited via collaborative tools like Google Docs and Overleaf.

7.3.2   The Four-Institute Structure

A structure with three institutes was kept until spring 2018, when the decision of aligning the number of MS programs with the number of institutes was taken, in particular with the arrival of the new provost, Sergey Masyagin, who replaced Tanya

Fig. 7.1 Faculty in December 2017

Stanko. Therefore, the Institute of Information Security and Cyberphysical Systems was created.

Since 2018, the organization in institutes of the university comprehends:

- Institute of Software Development and Engineering (initial and current Director Manuel Mazzara)
- Institute of Data Science and Artificial Intelligence (initial and current Director Adil Khan)
- Institute of Robotics and Computer Vision (initial and current Director Alexandr Klimchik)
- Institute of Information Security and Cyberphysical Systems (initial Director Alberto Sillitti and then Rasheed Hussain)

The four institutes are reporting to the Faculty of Computer Science and the dean's office: Giancarlo Succi was appointed in 2016.

Institute of Software Development and Engineering

This unit is organized in three labs:

- Lab of Operating Systems, Programming Languages, and Compilers—Head Eugene Zouev
- Lab of Software and Service Engineering—Heads in time order: Bertrand Meyer, Manuel Mazzara, and Nikolay Shilov
- Lab of Industrializing Software Production—Head Giancarlo Succi

Institute of Data Science and Artificial Intelligence

This unit is organized in four labs:

- Lab of Data Analysis and Bioinformatics—Head Yaroslav Kholodov
- Lab of Artificial Intelligence in Game Development—Head Joseph Alexander Brown
- Lab of Data Analysis and Machine Learning in the Oil and Gas Industry—Head currently to be appointed
- Lab of Machine Learning and Knowledge Representation—Head Adil Mehmood Khan

Institute of Information Security and Cyberphysical Systems

This unit is organized in two labs:

- Lab of Cyberphysical Systems—Head Yaroslav Kholodov
- Lab of Networks and Blockchain—Head Rasheed Hussein

Institute of Robotics and Computer Vision

This unit includes at the moment the Laboratory of Intelligent Robotics Systems, with head currently to be appointed.

7.3.3 Planning the Work of the Faculty

As the faculty was made of institutes and the institutes were made of labs, there was a key need to put all these elements together and for each faculty member to plan his/her own future, the one of his/her lab, and summing up all to institutes and faculty. For this reason, from 2018, new documents supplemented the overall mission document of the faculty: the manifestos of the institutes, which formed a key component not only in planning the achievements but in the real formulation of the faculty as a body of world-class researchers.

In parallel, again with the approach of having the documents growing bottom-up, specific teaching regulations were drafted, and a faculty handbook was produced. The handbook guides the life of faculty members, from appointment to promotion, up to retirement and awarding of emeritus status. Again, the content was not particularly different from the content of most of such documents worldwide, but having the individual professors sitting together and discussing literally word by word such content was a major propulsion in further enhancing the Innopolis spirit.

7.4 Faculty of Humanities and Social Sciences

In a second moment, a Faculty of Humanities and Social Sciences was created with Head Oksana Zhirosh. The faculty is composed of one institute, the Institute of Humanities and Social Sciences, which is composed of two labs: Humanities Laboratory and Social Science Laboratory. The idea of the institute is providing research and teaching in non-IT disciplines like pedagogy, sociology, English, and academic writing. Recently, such faculty has provided also support in areas that bridges the traditional disciplines of philosophy to computer science, such as logic, cognition, and artificial intelligence.

Acknowledgement The authors would like to thank Evgenii Bobrov for helping with this chapter.

Chapter 8
Cooperation with Industry

Fundamental research without applied research cannot have a direct impact on the economic ecosystem. Modern universities require intense cooperation with industry. The model of the Innopolis system is peculiar, companies support scholarships, and students have free education that they will pay back by working for 1 year after graduation.

8.1 The Environment

Innopolis University targets excellence in teaching IT and leading technological transformation in Russia. Digital technologies and IT sector permanently evolve as shown by dramatic changes in the recent years. Research and education programs have to be in constant contact with industry in order to preserve the relevance and value. Among other methods, consulting by faculty members is a natural way to be emerged in to the IT sector everyday problems and track the validity of research and education programs. Moreover, the consulting activities by faculty members convey a positive message to the industry with regard to the Innopolis brand and excellence, thus reinforcing the impact of the IU on the IT sector and society in general while contributing to a self-sustained university as a private education institution.

The major goals in this field can be identified as:

1. to become trusted advisor for the leading IT companies across Russia and Tatarstan
2. to establish proved track of records of projects delivered for leading IT companies in Russia and Tatarstan
3. to provide the mechanism for faculty members to establish direct touch with industry needs
4. to get additional funding for the IU to become a financially viable institution
5. to get additional funding for the faculty members to run research activities

© The Author(s) 2022
M. Mazzara et al., *Innopolis University - From Zero to Hero*,
https://doi.org/10.1007/978-3-030-98599-8_8

8.2   The Expertise

Due to their background and experience, IU faculty members have a natural capability to create and deliver dedicated trainings. These trainings may lead to certification such as Project Management Professionals or others. The faculty at IU have a strong experience in research activities, which in most cases implies their keen ability in developing prototypes and conducting studies. This may be particularly valuable for industry seeking to validate their hypothesis with scientific methods including prototyping.

IU is a hub where specialists of different IT-related disciplines and sectors meet. Thus, IU has a first-level access to professionals who are enthusiastic to participate in studies on technology trends in IT. In addition, faculty members are exposed to the cutting-edge technologies and are experienced in analysis methods. With that, IU can offer a unique view on the trends with a scientifically proven analysis.

Faculty members are knowledgeable in many very advanced areas of IT. They are also hands-on and operational to run short-term projects with high impact. Members are willing to participate to those industry projects within their capacity and in an area of their research interests. Nevertheless, the following limitations should be noted:

1. Time budget. During the teaching semester, the capacity of faculty is very low since their priority is teaching, though some would be able to devote from 1 h to 1 day per week to an interesting project;
2. Research interest. Faculty members have usually a very focused area of research interests that might be difficult to match with an industry demand.

A number of specialized courses can be taught by the faculty. Courses can be organized in different groups:

- **Courses derived from academic teaching:** these courses span from the basic to the advanced ones and are adapted versions of the current courses taught at BSc and MSc levels;
- **Courses derived from research activities:** these courses are generally advanced ones that can be created based on the expertise of specific faculty members and connected to past and current research activities;
- **Custom courses:** these courses can be designed ad hoc to satisfy the needs of a specific company (or a set of companies) focusing on specific topics and not present in any catalogue of the standard courses offered.

The faculty can also provide support to companies in the development of the application for a patent and in the management of the IP-related issues. Such activities may include:

- **Analysis of the patentability:** technical analysis of the invention to understand if it is possible to apply for a patent;
- **Plagiarism analysis:** analysis of the implementation of a technology to understand if it infringes a specific patent or copyright.

8.3 Experience of Corporate Training

Innopolis University team is specialized in delivering corporate training for management and developers. Several training sessions have been conducted for managers and developers of phone service providers and other business fields too. Typically, the sessions last for several days [10].

For example, only in 2018, more than 400 h of training were conducted involving more than 500 employees in 4 international companies. Although we cannot share the details of the companies involved, they are mid to large size and employ more than 10k people.

Figure 8.1 depicts one of the business trips for consultancy to one of the major Russian communication companies.

The trainings are typically focusing on:

- Agile methods and their application
- DevOps philosophy, approach, and tools
- Microservices

8.3.1 Organization and Delivery

The target group is generally a team (or multiple teams) of developers, testers, and often mid-management. We also suggest companies to include representatives from businesses and technical analysts and marketing and security departments. The nature of the delivery depends on the target group: sessions for management focus more on effective team building and establishment of processes. When the audience is a technical team, the focus goes more on tools and effective collaboration within and across the teams.

8.3.2 Structure

The events are typically organized in several sessions run over a 1-day to 3-day format made of frontal presentations and practical sessions. The sessions are generally conducted at the office of the customer in a space suitably arranged after the previous discussion with the local management. Whenever possible, the agenda and schedule of the activities have to be shared in advance. In this way, the participants know what to expect, and sometime a preparatory work is required.

Fig. 8.1 Business trip for consultancy. On the left Evgenii Bobrov, current provost for education

Acknowledgement The authors would like to thank Evgenii Bobrov and Alexandr Klimchik for helping with this chapter.

Part IV
The Future

In the previous parts, we have described the beginning of the enterprise and its early development, up to the present. In this part, we look at the future, how to internationalize the university and what strategies to follow to grow further. We will also look at the changes in education brought into the game by the pandemic of COVID-19. Success is ahead.

Chapter 9
Internationalization of the University

These days, universities acquire a reputation appearing on international rankings such as THE and QS. Innopolis is in the process of entering such rankings. In this chapter, we discuss how this challenge should be attacked and how the related processes should be organized in order to achieve the result. We also present some of the early successes of the university in this area. The chapter also analyzes the role of international conferences in achieving global reputation.

9.1 Internationalization and Globalization

Although the terms "internationalization" and "globalization" appear to be linked, they should not be used interchangeably. The former is defined as a process that "involves increasing the range of international activities within universities and between universities and other educational institutions and the numbers of international students and academic staff" [12], whereas the latter is defined as "trends of increasing international interdependence and growth of cross-border activities".

Globalization is pressuring universities around the world to reinvent themselves and implement transformative institutional changes that will serve as a "foundation for a balanced and integrated university experience at the interface of global and local exposure" [6]. Internationalization's transformative institutional changes are frequently seen as a tool to acquire a competitive advantage on international markets, as well as to internationalize university's current practices and strategies [12].

© The Author(s) 2022
M. Mazzara et al., *Innopolis University - From Zero to Hero*,
https://doi.org/10.1007/978-3-030-98599-8_9

9.2   The Ecosystem

The phenomenon of globalization leads these days to the creation of an unified educational space and market dominated by international rankings. Universities compete for *internationalization* with associated risks and benefits [7]. Russian education has a recognized reputation around the world, especially in fundamental science, which presents itself as a good selling point on international market. In 2012, the Russian government proclaimed *"internationalization"* to be one of the major objectives of higher education development.[1] To support the idea, the government launched a so-called "5top100" project,[2] which is aimed at "maximizing the competitive position of a group of leading Russian universities in the global research and education market". To put it in a simpler way, a group of leading Russian universities started to receive additional funding in order to facilitate their international attributes, particularly, increase positions held by this group of universities in the world university rankings (QS, *Times Higher Education*, Academic Ranking of World Universities).

Innopolis University (IU) is a young and ambitious university in Tatarstan in the Russian Federation, which has a strong focus on education and scientific research in the field of IT and Robotics [9]. It is located in the newly created Innopolis city (near the capital city Kazan) which also comprises ICT companies and the Innopolis Special Economic Zone. Innopolis aims to be the major Russian IT hub. In its development, the university was trying to follow the main trends of IT education borrowed from the world's leading higher education institutions. One of these trends was and still is internationalization. Since the very foundation, the university tried to hunt international faculty members as well as attracting international students. Additionally, the university has been developing various international initiatives including summer schools, exchange programs, conferences, etc. Existing for 10 years, the university succeeded in creating an international environment on campus.

9.3   Internationalization Strategy of IU

From the first step, IU showed its internationalization invitations. In December 2012, just few days after its creation, the university concluded an agreement with the Carnegie Mellon University (CMU). Under the agreement, the sides agreed to transfer a master's degree program on Software Engineering from CMU to IU. In spite of the fact that IU transferred its first educational program from CMU, IU always took as a role model a slightly different type of higher education institutes, i.e., young technical universities. The best examples worldwide are KAIST (South

---

[1] http://www.kremlin.ru/acts/bank/35263.

[2] https://5top100.ru/en/.

Korea), the IT University of Copenhagen (Denmark), and the Hong Kong University of Science and Technology (Hong Kong).

Innopolis University believes that internationalization will help in achieving the following goals:

- Improvement of educational and research quality;
- Focus on innovative and emerging technologies (i.e., improved "Agile" software development);
- Inclusion in many core courses of "real-life"-style production cycle exercises including a final product delivery (i.e., course projects);
- Competitiveness of IU's graduates and research results in international academic and IT market;
- Guaranteeing to IU a place as an equal participant of international cooperation.

To achieve these aims, IU set its main directions of internationalization that have been developed for the last 5 years:

- Recruitment of foreign students and faculty members;
- Employment of foreign faculty members;
- a scheme of visiting professorship from foreign institution;
- Mobility programs (student exchange; Erasmus+; research internships; joint PhD supervision);
- International olympiads and competitions;
- International conferences;
- Joint research and publications.

Despite the fact that IU is not part of 5top100 project, it does not mean that it has ignored world university rankings. In fact, IU valued them high and believed they potentially might help with the following issues:

- proving IU's academic and employment reputation;
- raise awareness about IU and Innopolis projects;
- requirement of international students;
- requirement of international faculty

For a young university, entrance into rankings is a fascinating challenge, and IU already makes its first positive steps in rankings. In 2018, it entered Round University Ranking,[3] where it reached a high score in internationalization indicators. It received maximum possible points for "Share of international co-authored papers" becoming the best result among Russian universities.[4] Additionally, Round University Ranking has a ranking table evaluating universities according "International Diversity", where IU came the second best among all Russian universities with only Lomonosov Moscow State University ahead. This particular ranking includes the following indicators: Share of international academic staff, Share of international

---

[3] http://roundranking.com/.

[4] http://roundranking.com/universities/innopolis-university.html.

Fig. 9.1 Official visit to the University of Luxembourg in April 2019

students, Share of international co-authored papers, Reputation outside region, and International level.[5]

In order to develop contacts with key universities, an uncountable number of teleconferences and numerous business trips have been conducted to those partner universities that were considered strategical. Figure 9.1 shows the official business trip to the University of Luxembourg in April 2019, while Fig. 9.2 depicts the students fair we attended in the same city in November. The photo in Fig. 9.3 shows a business dinner during the visit to Windesheim University of Applied Sciences (with Iouri Kotorov, head of International Affairs in 2018–2020).

9.4   The International Conferences

Innopolis University organized a number of international conferences in order to get more visibility. Here we will just present some for which we were directly involved and are aware of the details. Of course, other colleagues organized other equally successful events. For example, at the end of the section, we will mention the experience of the Robotics Institute.

---

[5] http://roundranking.com/ranking/world-university-rankings.html#nternational-2018.

Fig. 9.2 Participation to the students fair of the University of Luxembourg in November 2019

9.4.1   The 10th International Ershov Informatics Conference (PSI 2015)

The first of all the conferences in the timeline was the 10th International Ershov Informatics Conference (PSI 2015). PSI conference is one of the most significant computer science conferences in Russia.

Fig. 9.3 A business dinner in October 2019, Zwolle in the Netherlands

The event featured David Parnas[6] as a keynote speaker. Parnas is pioneer of software engineering. His concept of information hiding in modular programming is an essential element of object-oriented programming.

The conference organization took 1 year, and it was not easy to make it happen given a lot of uncertainty and complicated logistics. It finally happened.

9.4.2   TOOLS 50+1: 2019

The TOOLS conference series is a long-running conferences on object technology, component-based development, model-based development, and other advanced software technologies. The name originally stood for "Technology of Object-Oriented Languages and Systems" although later it was usually no longer expanded, the conference being known simply as "the TOOLS conference". The conferences ran from 1988 to 2012 and were finally revived in 2019 in Innopolis under the name TOOLS 50+1 on October 15 to 17, 2019. Figure 9.4 shows the social event of the conference.

---

[6] https://en.wikipedia.org/wiki/David_Parnas.

Fig. 9.4 Social dinner of TOOLS 50+1

9.4.3 OSS 2020

The goal of the International Conference on Open Source Systems (OSS) is to be a holistic forum on Free/Libre Open Source Software (FLOSS) aimed across people from widely diverse backgrounds.

The edition 2020 was planned to be held at Innopolis University from May 12 to 14, 2020. Due to travel restriction amid the pandemic, the event was fully moved online.

The focus of the 2020 edition covered two major aspects:

1. The development of Free/Libre Open Source Systems and the underlying technical, social, and economic issue;
2. The adoption of FLOSS solutions and the implications of such adoption both in the public and in the private sector.

The fact that the event was run online helped even further in the dissemination of the concept and the visibility of the university.

9.4.4   Events in the Field of Robotics

The Robotics Institute has been very active since the early days in the organization of events. Notably the "Nonlinearity, Information and Robotics" conference was organized in 2020[7] and 2021,[8] and there are plans for the edition of 2022. The conference proceedings, indexed by Scopus, are published by IEEE.

Besides the conferences, a significant number of summer and project schools have been hosted by Innopolis University:

- Double Summer School on Collaborative Robotics and Software Engineering, August 6–19, 2018
- International Project School on Self-Driving Vehicles, April 6–13, 2019
- Double Summer School "Robot's Cognition, Perception and Control", June 10–23, 2019
- Summer School on Machine Learning in Robotics, December 7–10, 2020[9]

Acknowledgement The authors would like to thank Evgenii Bobrov, Petr Zhdanov, and Alexandr Klimchik for helping with this chapter.

---

[7] http://nir.innopolis.university/2020.

[8] http://nir.innopolis.university/2021.

[9] https://events.innopolis.university/winterschool.

Chapter 10
Online and Blended Education: After COVID-19

The year 2020 changed the panorama and horizon of teaching forever. Innopolis University switched to online and blended education with minor problems only. Why we were "almost" ready for the switch. This chapter presents those days, how we reacted and managed to move on. What is going to be the future of education?

10.1 Switching to Online Education

Life-changing events have occurred in the year 2020, affecting a wide range of professional fields. One of the most affected areas has been education. Institutions all across the world had to make the transition to online education in a matter of days.

With the onset of the Fourth Industrial Revolution, various aspects of the present educational system have shown to be outdated. Despite the fact that online education is here to stay, frontal classrooms are a millennia-old tradition that cannot be completely replaced without ignoring human nature. Old and new can coexist, and humans and machines can work together to advance society.

Figure 10.1 shows a pre-COVID classroom where students are raising their hands to get the attention of the teacher, a practice that is at least as old as the First Industrial Revolution.

© The Author(s) 2022
M. Mazzara et al., *Innopolis University - From Zero to Hero*,
https://doi.org/10.1007/978-3-030-98599-8_10

Fig. 10.1 Students asking for the attention of the teacher

10.2   Lessons Learned from COVID-19

The year 2020 has been a remarkable one for the world in general and, in our case, specifically for education. Many certainties vanished in a matter of days or weeks and will not be back. At Innopolis, we have been forced to reorganize the educational process in 1 week. This implied improving the ICT infrastructure, purchasing or extending licenses for specific platforms such as Microsoft Teams[1] or Zoom,[2] making sure that connectivity was suitable, and identifying adequate methods of instructions. Some challenges have been institutional, some individual. Even something as simple as setting up an effective home office with good connectivity has been, at times, challenging. The reluctance of teachers to indulge in the new format can also be a challenge. However, as an IT university, we did not experience this to a large extent. Changes came as a hurricane, and they are not leaving. Online education, in the blended format, is here to stay.

---

[1] https://www.microsoft.com/en-us/microsoft-365/microsoft-teams/group-chat-software.

[2] https://zoom.us/.

The experience of the year 2020 showed some distinctive emerging traits that we believe can be generalized to any educational organization worldwide.

10.2.1 New Methods of Teaching

Education in the form of online streaming has existed before as an integrative option for certain organizations. For some educational commercial online platforms (or universities) was instead a regular way of operating. Traditional academia was mostly reluctant to move and adopt it. The year 2020 was a game changer. During and after the pandemic, new methods of teaching are needed to facilitate student learning. The most popular method is the usage of videos, either the real-time record of a lecture or material separately prepared and provided to students. These options provide flexibility to students and the possibility to re-watch multiple times or even to watch in a speed mode. Recorded lectures are ways to accommodate the lack of physical contact. However, videos by themselves may hardly be the most effective methods of knowledge delivery, due to the limited interaction between students and professors or among students themselves [1]. Therefore, the instructional approaches shifting to online modality should be considered in the light of different factors. Teachers will be operating more as moderators than as instructors [5]. Flipped learning is an alternative method where a teacher records a video of the lecture and shares it with students to be watched as homework, before the lecture. At lecture time, the focus will be on the discussion of the content, on answering questions, and on clarifying misunderstanding [3].

10.2.2 Evaluation and Assessment by Virtual Means

Adapting courses and programs to an online format is obviously difficult, but the most difficult aspect is assessing students' results and proctoring. Assessments should be done on a regular basis during the semester rather than relying simply on final exams: the traditional final exam will be replaced by continuous evaluation. Many universities have already switched to this format, and the year 2020 accelerated this transition.

10.2.3 Greater Use of Open Educational Resources

This has already happened for some time; 2020 is a year of non-return. Here, we also advocate the importance of traditional content, such as books. Closed-access resources turned out to be a blockage for online education. Some of the advantages of online education are affordability and accessibility, and, to achieve this, open educational resources are fundamental.

10.2.4   Professional Development for Teaching Staff on Digital Education

Our organization, Innopolis University, is a young and dynamic IT university which experienced marginal issues in the switch, being the teaching staff mostly IT specialists with long experience as software users and developers. However, we noticed that the changes have been problematic for some of our colleagues from less IT-related departments, and we have observed significant problems in other universities and faculties. The path of complete digitalization for established teachers is long ahead.

10.2.5   Distinctive Features Must Compensate the Decrease of Campus Experience

Education is not only about content delivery but campus experience, human networking, sports, and social activities, where young adults learn a 360-degree perspective on life and profession. Online education, even in a hybrid format, cannot offer a comparable experience in this sense. We certainly hope that part of these offline activities can be eventually restored. However, it is necessary to rethink education to offer specific features that can compensate for this emerging gap.

10.2.6   Greater Emphasis Will Be Placed on Collaborative Projects

As a consequence of loss of campus interaction and networking, collaborative projects, even if executed remotely, can reinstate back some feeling of community belonging and horizontal learning.

10.2.7   ICT Infrastructure Is Critical

While in the pre-COVID classic delivery mode, teaching could be potentially delivered with a shortage of ICT infrastructure, now this is a "condicio sine qua non". ICT was before a support infrastructure, and every teacher was able occasionally to deliver a functioning class without a projector, a laptop, or an Internet connection. This is not possible now, and often even a slightly sub-optimal bandwidth can make things frustrating and difficult to follow. Universities that want to win the race have to put the development of ICT infrastructure on top of the list. Before considered a distraction by many lecturers, laptops in the classroom are now the primary tools of operations.

10.2.8 Paradigm Shift in Teachers' Training Programs

Teachers' training programs are mostly designed for classroom-based in-person teaching. With online education as a norm, teachers have to be trained accordingly. Online education presents more challenges in terms of ensuring students' attention. Furthermore, classroom interactions allow teachers to have some idea of students' mental state, and teachers can provide support accordingly. Online education makes it difficult to know students' issues (especially with their cameras and microphone off most of the time). It might become mandatory for teachers to have basic mental health education to assess students' behavior and provide support accordingly.

10.2.9 New Ways of Establishing Discipline and Roles

With education being learner centric, classroom disciplinary restrictions are more relaxed as compared to the situation some decades ago. Students have a say in educational reforms, and any significant alteration in the educational process takes place with the teacher and student consensus. With online education, classroom boundaries need to be revisited in consultation with teachers and students. By boundaries, we refer to teacher-student agreements that ensure discipline required to maintain quality of education. Some of the points that require consideration involve camera and microphone on/off issues and lecture delivery modes such as recorded or live sessions.

10.3 Final Thoughts

Education is not only about content delivery and even not principally about it. What creates the unique university experience is human networking, sports, and social activities, all the involvements where young adults learn a 360-degree perspective on life and profession guided by senior colleagues and faculty. Online education, even in a blended format, cannot offer this experience. There is a significant risk that students can escape "reality" and hide in a "cocoon". After decades of exaggerated emphasis on "soft skill" where introverts were pushed to their limits and extroverts could more easily thrive, we now push young generations to the opposite extreme and make introverts thrive. Even if one could see a sort of "divine justice" in such change, it is not by moving from one extreme to another that we balance a situation. In some sense, injustice cannot compensate for another injustice. We need to have a balanced approach. Work is now necessary to find a way to overcome such a possible degeneration and avoid the *"student in a learning cocoon"* approach. Even in the post-2020 era, the community still counts.

The future will see a sharp separation between those educational institutions able to catch up with the pace and those left behind, both in terms of new pedagogical methods and supporting ICT infrastructure. In the pre-COVID classic delivery mode, teaching could be delivered with a modest ICT infrastructure: projectors and laptops were often sufficient, and an Internet connection is not always necessary. Every teacher was able to deliver functioning classes without the need for any particular device. Now even a slightly sub-optimal bandwidth can make things impossible.

Chapter 11
Faculty Strategy for the Future

For the future, the vision is transforming the university into a multidimensional center of excellence, still focused on IT, but not just related to pure computer science, including possible synergistic domains of knowledge.

For us, **excellence** is the worldwide recognition of competence in the target research areas, the attractiveness of excellent students ready to invest their life and their wealth in the educational programs, and the high demand of research and consulting support from the industry.

11.1 The Goals

From the high-level vision, in order to make plans implementable, it is necessary to identify more concrete low-level goals. We identified the following:

1. Join the top 100 university in QS
2. Educate outstanding professionals
3. Aim at self-sufficiency

11.2 From Key Objectives to Overall Goals

To move toward the identified goals, we identified specific objectives. By achieving such objectives, we are convinced that we will make a significant move ahead.

© The Author(s) 2022
M. Mazzara et al., *Innopolis University - From Zero to Hero*,
https://doi.org/10.1007/978-3-030-98599-8_11

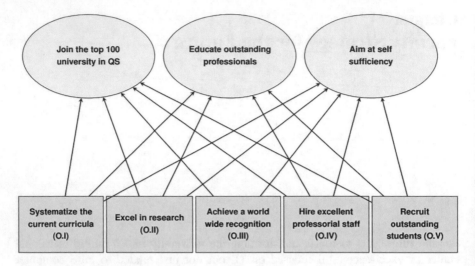

Fig. 11.1 From key objectives to overall goals

O.I: systematize and strengthen existing curricula at the BS, MS, and PhD level,
       considering possibilities for their expansions in two main dimensions:

- new subject areas;
- tailored teaching schema for excellent and entrepreneurship-oriented
  students.

O.II: leverage already strong research competences to excel, also trying to build
       stronger relationships with local and international granting agencies and
       industrial partners.

O.III: achieve a worldwide recognition.

O.IV: hire excellent professorial staff, existing or potential worldwide leaders in
       the field, also using the mechanisms of visiting and adjunct professorships, a
       prerequisite for any expansion of the educational offer and further research
       endeavors.

O.V: recruit nationally and internationally outstanding students at all levels, from
       BS to PostDoc.

The relationship between the key objectives and the overall goals is represented
in Fig. 11.1.

11.3   Means to Achieve the Objectives

To achieve the long-term goals via the stated objectives, the faculty has identified
the following means:

M.1: organization of the operations in institutes and laboratories centered on:

 M.1.1: focused research subjects,
 M.1.2: core educational areas and programs,
 M.1.3: targeted consulting infrastructure

M.2: suitable handling of the professorial profiles and careers,
M.3: dean's office,
M.4: strong synergy with the international department and the student recruitment department,
M.5: effective budgeting,
M.6: advisory board,
M.7: specific mechanisms to enact via the various bodies of the faculty.

The first means to achieve the objectives of the faculty is the organization of its operations, with particular attention to:

- the research performed through institutes and laboratories,
- the core educational areas and programs,
- the consulting offering.

11.4   Faculty Operations

An effective management of the activities of the faculty is an essential ingredient for the overall success of the academia. Therefore, the faculty operations have been organized in three levels:

- overall university academic council
- faculty cabinet
- dean's office

11.5   Strong Synergy with the International Department and the Student Recruitment Department

Especially with reference with the objectives of achieving a worldwide recognition (11.2) and of recruiting nationally and internationally outstanding students at all levels, from BS to PostDoc (11.2), the activities of the faculty proceed hand in hand with two core departments of the university: the international department and the student recruitment department.

11.6 Effective Budgeting

The implementation of the plan depends on the resources that have been allocated to it; an effective allocation of resources is therefore a prerequisite for an efficient implementation.

In budgeting the resources, therefore, on the one side, direct responsibility is given to the institute directors and the lab heads, and, on the other, the distribution is tight to the level of contribution to the overall goals of the university.

11.7 Advisory Board

To provide a long-term guidance of the activity of the faculty, it would be useful to create a body of outstanding, worldwide researchers that could advice the faculty as a whole, the dean, and the vice-provost for education on a variety of issues, including the definition of the goals, the structure of the curricula, the recruitment of new research and professorial staff, and the career evolution of current staff.

Chapter 12
Conclusion

Our journey over the years has been accompanied by the majestic sunsets over the Sviyaga River that are visible from the university building (Fig. 12.1). Lots of sweet memories are hosted in our hearts. The university is growing and new colleagues are joining. We knew and hoped that this was going to be the future. It is now here.

At the same time, our graduates are occupying important positions in companies and universities in Russia and abroad. Several of them also started up business that is growing and becoming relevant. Figures 12.2 and 12.3 show the graduation ceremony of August 2021, which happened while this book was still in the writing phase.

Life's developments always have sweet and sour notes; something is always waiting for us after the next river turn. We need to keep sailing to see what is there. We do not have bird's eyes on life. Whatever will happen and wherever we will be, the physical separation will not affect the spiritual bond created over the years.

Stay blessed, our dear companions of this journey.

© The Author(s) 2022
M. Mazzara et al., *Innopolis University - From Zero to Hero*,
https://doi.org/10.1007/978-3-030-98599-8_12

Fig. 12.1 Winter sunset

Fig. 12.2 Graduation, summer 2021

Fig. 12.3 Manuel Mazzara at graduation, summer 2021

Part V
A Visual Journey

A large number of pictures have been collected over the years; some of them have been already added to the first four parts of the book. The fifth part is special; text is serving pictures and not the other way around. It is a visual journey into the Innopolis development. This part has been built on top of the beautiful photo archive of our colleague Timur Tsiunchuk. Sometimes an image is more informational than hundred pages.

Chapter 13
Photobook

13.1 Prof. Meyer and the Office in Kazan

Figure 13.1 shows Prof. Bertrand Meyer teaching a class in a lecture room of the office in Kazan in the academic year 2014–2015.

Prof. Meyer was one of the first contributors of the project. He was regularly visiting our small offices and helping with curricula design, knowledge transfer, and class delivery. Bertrand is not only a good friend but a key player in the development of the university. He is still a visiting professor and comes often to Innopolis.

13.2 The Inauguration of Innopolis City

Figure 13.2 presents the rector of the university, Alexander Tormasov, during the inauguration of the city and the university building at the presence of the Prime Minister Dmitry Medvedev on June 9, 2015. That day was a big one with a large number of people coming from all over Russia. Personalities, media, and journalists: everyone wanted to be there! We were there.

13.3 The Trip of Students to Shenzhen

Figure 13.3 shows a delegation of our first cohort of bachelor students visiting Huawei in Shenzhen (China) in summer 2015, in a leisure moment. The students were directly invited by the company to their main site. Huawei kindly offered to

© The Author(s) 2022
M. Mazzara et al., *Innopolis University - From Zero to Hero*,
https://doi.org/10.1007/978-3-030-98599-8_13

cover the costs and provided technical education to the students for a period of 2 weeks. They come back to Kazan with excellent memories and nice book about Shenzhen.

13.4   The Faculty Retreat

In Fig. 13.4, bright memories of a faculty retreat in Italy are shown. The faculty (or part of it) self-organized a trip to the north-west of Italy (Liguria and Piemonte) in summer 2016. The idea was to mix work, reflection, and leisure. A similar concept was then repeated in 2017. When the numbers grew, the logistics become more complex, and the retreats were usually hold in Tatarstan.

13.5   The First Graduation

During the same summer of 2016, at the end of August, Innopolis had the first graduation of master students, depicted in Fig. 13.5. Here Joseph Brown is dressing the gown of his alma mater, the University of Guelph, Canada. Using the gown (academic regalia) of the alma mater at student graduation is a common tradition in many countries beyond the UK these days. This practice dates back to Oxford and Cambridge. Dr. Brown brought it into our environment since early on.

13.6   The Lecture Theaters

Figure 13.6 shows one of the authors, Manuel Mazzara, in one of the large lecture theaters of Innopolis University. The excellent infrastructure of the university has been one of the important enablers for the growth of the entire city.

13.7   A Success Story, Among Many

Some of our graduates followed the academic path, doing PhD studies all over the world, including places as Cambridge. Many landed in very good positions in top companies as software engineers, in Russia and abroad. Others became entrepreneurs. Among the several notable success stories of our graduates, one is standing out.

Kevin Khanda (in Fig. 13.7) was enrolled at the university in 2015 for his first year of bachelor studies and graduated in 2019. Since the early days of the program, he had the great ambition to create a competitive company, to do something new and big, and to work on some extraordinary project. And so he did, launching KazanExpress in 2017 and attracting investors. KazanExpress is a resident company of Innopolis city, and it is the first marketplace able to deliver orders in 1 day. Since 2018, the number of orders kept increasing to the point that, in 2021, AliExpress Russia became an investor.

Fig. 13.1 Lecture by Bertrand Meyer in the initial operational center of Kazan

Fig. 13.2   Inauguration of the city

Fig. 13.3   A leisure moment in China

Fig. 13.4 Rest and work in Italy

Fig. 13.5 Joseph Brown, the first master graduation in the university building, 2016. On the left, Ales Zivkovic; on the back, Rasheed Hussain

Fig. 13.6 Manuel Mazzara in 2018—"© 2018 Sasha Malikov, all rights reserved, printed with permission"

Fig. 13.7 Kevin Khanda and Giancarlo Succi at the graduation

Acknowledgement   The authors would like to thank Timur Tsiunchuk for providing some of the photos appearing in this chapter.

References

1. Sun A. and Chen X. Online education and its effective practice: A research review. journal of information technology education. *Journal of Information Technology Education*, pages 157–190, 2016.
2. Roman Bauer, Lukas Breitwieser, Alberto Di Meglio, Leonard Johard, Marcus Kaiser, Marco Manca, Manuel Mazzara, Fons Rademakers, Max Talanov, and Alexander Dmitrievich Tchitchigin. The biodynamo project: Experience report. In *Advanced Research on Biologically Inspired Cognitive Architectures*, pages 117–125. IGI Global, 2017.
3. J.L. Bishop and M.A. Verleger. The flipped classroom: A survey of the reaserch. *Paper presented at the ASEE National Conference Proceedings, Atlanta, GA*, page p.5, 2013.
4. Lukas Breitwieser, Roman Bauer, Alberto Di Meglio, Leonard Johard, Marcus Kaiser, Marco Manca, Manuel Mazzara, Fons Rademakers, and Max Talanov. The biodynamo project: Creating a platform for large-scale reproducible biological simulations. *4th Workshop on Sustainable Software for Science: Practice and Experiences (WSSSPE4)*, 2016.
5. Michael B Cahapay. Rethinking education in the new normal post-covid-19 era: A curriculum studies perspective. *Aquademia*, 4(2):ep20018, 2020.
6. Michael Cross, Ehpraim Mhlanga, and Emmanuel Ojo. Emerging concept of internationalisation in south african higher education: Conversations on local and global exposure at the university of the witwatersrand (wits). *Journal of studies in international education*, 15(1):75–92, 2011.
7. Yuan Gao, Chi Baik, and Sophia Arkoudis. *Internationalization of Higher Education*, pages 300–320. Palgrave Macmillan UK, London, 2015.
8. Tahira Jibeen and Masha Asad Khan. Internationalization of higher education: Potential benefits and costs. *International Journal of Evaluation and Research in Education (IJERE)*, 4:196–199, December 2015.
9. D. Kondratyev, A. Tormasov, T. Stanko, R. C. Jones, and G. Taran. Innopolis university-a new it resource for russia. In *2013 International Conference on Interactive Collaborative Learning (ICL)*, pages 841–848, Sept 2013.
10. Manuel Mazzara, Alexandr Naumchev, Larisa Safina, Alberto Sillitti, and Konstantin Urysov. Teaching devops in corporate environments - an experience report. In *Software Engineering Aspects of Continuous Development and New Paradigms of Software Production and Deployment - First International Workshop, DEVOPS 2018, Chateau de Villebrumier, France, March 5–6, 2018, Revised Selected Papers*, pages 100–111, 2018.
11. Bertrand Meyer. Applying "design by contract". *Computer*, 25(10):40–51, October 1992.
12. Sue Robson. Internationalization: A transformative agenda for higher education? *Teachers and teaching*, 17(6):619–630, 2011.

Printed in the United States
by Baker & Taylor Publisher Services